ARCHITECTURAL
DRAWING

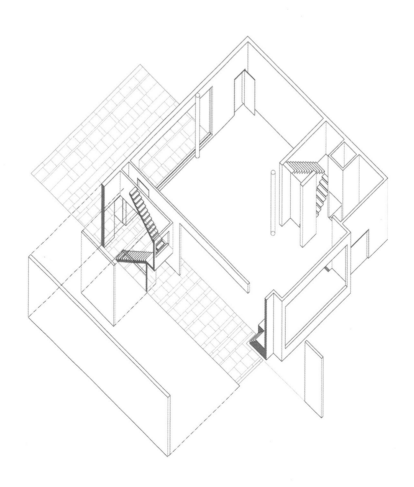

Published in 2010
by Laurence King Publishing Ltd
361–373 City Road
London EC1V 1LR
Tel +44 20 7841 6900
Fax +44 20 7841 6910
E enquiries@laurenceking.com
www.laurenceking.com

A catalogue record for this book is available from the British Library

ISBN 978 185669 679 1
Series design by John Round
Book designed by The Urban Ant Ltd.
Printed in China

DAVID DERNIE

ARCHITECTURAL DRAWING

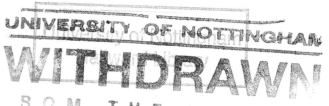
Laurence King Publishing

Contents

Related study material is available on the Laurence King website at
www.laurenceking.com

Introduction

'What I believe is that whether it be a question of sculpture or of painting, it is in fact only drawing that counts. One must cling solely, exclusively to drawing. If one could master drawing, all the rest would be possible.' **Alberto Giacometti**

This book celebrates the wide range of drawing techniques now available to architects. It looks at conventional and less conventional drawings and the methods used to make them in an attempt to open up creative approaches to architectural visualization. At a time when buildings and components can be wholly manufactured digitally, this book attempts to readdress the whole question of drawing as a way of thinking, a notion that is common in other visual arts.

Digital media now offers unprecedented opportunities for architectural drawing, but at the same time one instinctively feels that there is an important dialogue to be maintained with other kinds of drawings and techniques that may reflect different kinds of architecture, imaginations and processes. This book explores a diverse range of drawing types, emphasizing the role of drawings as vehicles for 'thinking about', rather than simply 'illustrating' architecture. Using examples from fine art, photography and stage design, the text explores the interdisciplinary nature of modern drawing, its integration with digital making, and the role the act of drawing can play in the exploration of the spatial and material conditions of a situated event.

The approach here is to underline the complementary relationship between traditional techniques and computer-generated images. Celebrating this dialogue allows us to explore the 'drawing as artifact' where the boundaries between drawing and modelling are eroded. The processes of drawing and modeling both mediate between our imaginations and the physical reality of architecture.

Drawings are the first steps in the process of making and as such it is important to recognize the relationship between hand and digital drawing, just as we retain the value of traditional and digital manufacture. By opening up the scope of architectural drawing, to techniques most often used by disciplines outside of architecture, we are able to explore ways of describing more diverse ideas and more varied ways of making architecture.

The engagement of drawing with materials and physical space is illustrated in a number of other visual arts. Here we will briefly illustrate the work of the photographer George Rousse, who explores the tangible concreteness of drawing onto found space and then the stage designer Caspar Neher, Brecht's stage set designer, whose drawings lie between the material and the imaginary and, like architecture, are focused on the framing of a human drama.

This kind of excursion into disciplines that lie on the edges of architecture is illuminating not only in terms of technique, but also in terms of the ways we think about drawings and the ways in which they are presented. The challenge for the architect is to find appropriate technique that engages with individual design direction and also to make the right types of drawings that take intentions forward and communicate them clearly. We must recognize that the process of visualization of architectural intent or experience is as complex and it needs to be subtle: compare, for instance, a street plan or section with the experience of street life. Often multi-sensorial experience of architectural spaces cannot be fully expressed by one drawing type or a singular drawing technique. Rather, it is only through a combination of drawings, collage, animation or film that the richness of architectural experience can be fully explored.

In other visual fields the importance of drawing and 'mark making' is increasingly recognized in relation to the spontaneity of creative thinking but it is perhaps not surprising, given the nature of the discipline, that the immediacy of hand-made drawing has all but vanished from today's architectural profession. During a relatively short period the 'hand-drawn' in architecture has, for the most part, been eroded first by mechanical and then later, digital, devices. Now the 'hand-drawn' image has been supplanted by the 'plotted', or indeed 'calculated image', and the physical 'act of drawing' has been reduced to a more passive mediation with a screen and its peripherals.

At best, however, these advanced techniques of representation enable us to explore form and to render light, colour and texture with a breath-taking precision. Advances in software and interface design now allow digital drawing a freedom of expression hitherto only available by hand: the best of this creative digital representation is distinctive, provocative and revealing.

Such expressive, digital drawings complement hand drawings: both are means to articulate ways of thinking and both can, at best, be immediate, individual and synthetic. Whilst it may be argued that hand drawings, by virtue of the directness of exchange, exclusively 'lay bare thought', it is also true that once mastered, digital drawing establishes its own realm of expression: both sets of tools can be considered means to 'materialize thought'.

Right
Troyes (1986) by Georges Rousse.

Historically, architectural drawings echoed the traditional craft of making buildings: Waterhouse's watercolours of masonry façades or Horta's intense pen and ink sketches for interior details, for instance, expressed a deep understanding of fabrication. In a similar way, Carlo Scarpa's pencil and crayon drawings speak of his practical knowledge, and in their hesitancy and incompleteness they may also reflect his dependence on the experience of his close circle of craftsmen and artisan friends (below).

Today, construction processes have become more varied and more integrated with digital information and intelligent manufacturing processes. Digital technologies now facilitate every stage of the design process through to manufacture. As product and construction processes increasingly take up methods of digital production, a new generation of architects is exploring the language of 'digital architecture'. In the same way that the 'hand of the architect' is arguably visible in traditional techniques, this new generation of architects is able to find its voice through refined understanding of digital drawing techniques, that can be rapidly modeled, manufactured and ultimately assembled on site.

In this perspective the material imagination (which Gaston Bachelard once famously described as an 'amazing need for participation which, going beyond the attraction of the imagination of forms, thinks matter, dreams in it, lives in it or in other words materializes the imaginary') underpins a creative digital drawing process, that moves from geometric description and spatial forms, to rendering of light and material textures that describe scale and tactile qualities. In such a process drawings and models hold the key to the coherence and integrity of the design, as they alone can articulate a relationship between the theme or intention of a project and its material form.

Mixed media drawings (drawings that are made using a mixture of hand and digital tools) can also be very useful to open up questions of intention, form and material. One such drawing *Imaginary Cities (1)* (page 9) uses a wide range of materials (from resin and bitumen to canvas and jute), to create a materially rich surface intended to articulate themes relating to memory, the history of the site and its people. It is intended as a sketch study and focuses on the spontaneous level of creativity as a preparatory stage of design. As Dalibor Vesely has described, the work is 'defined by the intention to return to the stage of design where the first attempt to visualize the content of design is taking place. The process of visualization can be described as materialization, or more precisely as the first encounter with the material conditions of the later, more abstract stages of design'.[1]

In a different way, and at a later design stage, Steven Holl's delicate watercolour of the Chapel of St Ignatius (1997) has a wonderful quality that communicates the character of colour and light that the eventual building was to embody. The drawing explores formal and material ideas and uses the unique qualities of watercolour to achieve a glowing

Right
Preparatory drawing for the Castelvecchio Museum, Verona by Carlo Scarpa.

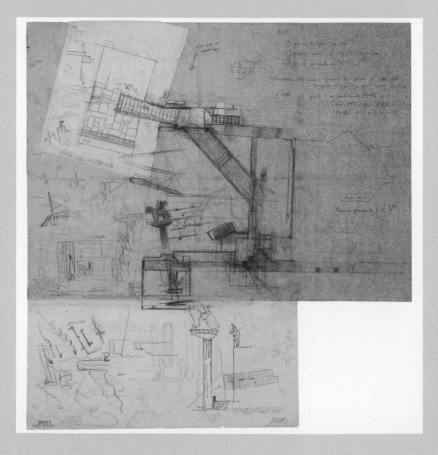

transparency to the overlapping 'bottles of light' (page 11).

Such drawings may be described as 'material drawings', exploring the physical character of the drawing's surface creating a reference to the actual material, rather than acting as straightforward illustrations of the materials to be used. A 'material drawing' integrates material and formal ideas into one texturally rich surface in order to maintain a synthesis of these dual horizons during the design process. Taken further, this work would be related to some of the concerns of the French photographer, Georges Rousse, whose work with existing buildings is inspiring for architects.

Drawn to the physical presence of dilapidated architectural space and deriving some impetus from the work of Gordon Matta-Clark in the 1960s and 1970s, Rousse started to work with drawings, paintings and photography in abandoned buildings during the 1970s. But while Matta-Clark largely carried out real scale 'photographic operations' in the buildings, Rousse draws, paints, cuts through and builds into the existing space to uncover new or underlying orders to the location (page 7).

The photographs represent a tension between the imaginary and the primary material order of the given space, discovered through drawing and its extension into a range of painting and photographic techniques. The work explores drawing at different scales that grow out of a response to a physical engagement with the place. The lines or hatchings become material edges, boundaries or implicit structure within real space; delineators of a new, fictive topography that are juxtaposed like a collage: Rousse's work exposes what he refers to as a 'theatricalization of space', a process of 'laying-bare' something new about a place.

Most importantly, Rousse's work is a process of discovery, as opposed to an illustration of a concept, a character that John Berger describes as the lynchpin of what it means to draw: 'Nearly every artist can draw when he has made a discovery. But to draw in order to discover – that is the godlike process, that is to find effect and cause'.[2] As in the early stages of an architectural design, Rousse's drawing process discovers new relationships, in part inherent in the given space (or programme), and in part composed of a new agenda introduced into the space that is subsequently transformed or re-presented.

This process uncovers specific relationships between sculpture, light, scale and the material conditions of enclosure and brings together a synthetic experience of a given space. Developing this understanding of material and spatial experience further are the drawings of Brecht's well-known stage designer, Caspar Neher (1897–1962). For Neher, the physical boundaries of a space were always focused on the realization of the human setting, and in a way that is particularly instructive for an architect, Neher developed a single-minded focus on human drama as the content of the work.

Working mostly with inks, washes and pens, his

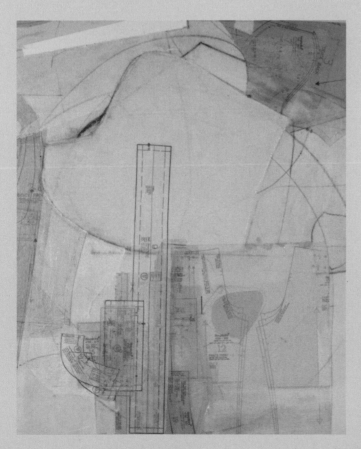

Left
Imaginary Cities 1, by David Dernie.

Above
Stage design for Brecht's *The Mother*, scene 2, by Caspar Neher.

drawings display a controlled sensitivity to tone and line, scale and lighting, that still today appear as though the words of the script were transposed into the drawing itself. Neher has been said to have written drama in the medium of drawing: 'Neher's sketches anticipated a production by a particular director with particular actors and a particular Ensemble. They were not interchangeable decorations for some production or other with conceivable alternatives. He was not sketching "stage pictures" but the play.'[3]

The freshness of these sketches, the way they simultaneously express material, light and scale of setting in so far as they are relevant to the play and the unfolding of the drama, are relevant to architecture where drawings rarely have such an open, synthetic and precise quality. Like architect's drawings, Neher's sketches were practical tools to convey information; Brecht would rely absolutely on their instruction. In this sense such arrangements of line and tone have an inherent precision, if this is defined as a reflection of thinking, rather than a corollary of completeness. Neher's sketches are more than illustrations of a preconceived idea, rather they have a spontaneity of an exploratory gesture that looks for the inspiration through the act of drawing. Like Rousse's investigations into space, they are drawings of discovery, but they are also drawings focused on situated human action.

Bringing together several of these themes are the extraordinary hybrid drawings of Sara Shafiei and Ben Cowd whose work is representative of a new generation of architects. Their studio attempts to move conventional architectural drawings, such as sections and plans, off the page, from two-dimensional surfaces to three-dimensional constructs. The purpose of the work is to re-define and exceed the traditional limits of drawing, using new technology such as laser cutting to layer, wrap, fold and use the inherent burn from the laser cutter to convey depth and craft. Their drawings, that are featured throughout the book, establish a tentative balance between ideas of craft whilst using newly established modes of design and technology and recognizing the intrinsic link of drawing to innovative manufacturing techniques, transforming paper into models.

These 'drawings as artefacts' are extraordinary concentrations of visual and creative experience, synthesized through the disciplined mastery of both traditional and digital technique. They represent a tradition of visual expression, where the reciprocity between thought and material are laid bare, as Yves Bonnefoy expresses: 'I have always understood drawing to be the materialization of the continually mutable process, the movements, rhythms, and partially comprehended ruminations of the mind.'[4]

In this sense, as we have seen, the fluidity and continuity of the drawing process is key, as an architect seeks to translate ideas into buildings. The process of drawing, as a process of 'materializing thoughts', is a creative, and inevitably individual process, the diversity of which can be reflected in the range of drawing tools and drawing types

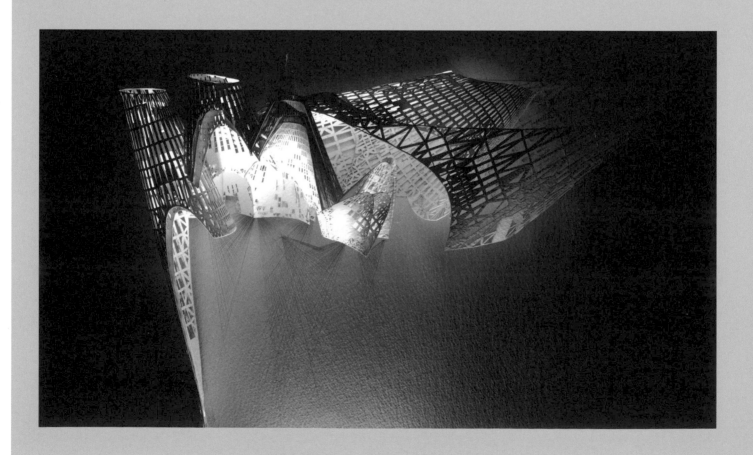

adopted. The following book is an attempt to capture some of today's drawing and representational techniques. Ranging from digital and computational drawings through to pencil and charcoal, the book is not exhaustive, but designed to offer insight into techniques that may enable individuals to find their own voice through the act of drawing and making.

About this book

The book is in three parts: Media, Types and Places. Media explores the tools used to make drawings; it takes the position that the computer is one of a number of tools that can be used for architectural drawings, in an attempt to encourage experimentation beyond predictable software products. It discusses line drawings, render and mixed media. The second part, Types, then describes the most common drawing projections used in architectural projects: these range from conventional projections to less conventional combinations of drawings. The final section, Places, describes three basic topographies that architectural drawings describe: interiors, landscapes and urban contexts. Each of these is illustrated with a variety of drawing types and media.

The book is intended to be both inspirational and practical. It is designed to encourage ambition and diversity in architectural drawing and, at the same time, to be a practical guide; a useful starting point, but not an exhaustive manual. A deeper understanding of drawing comes more directly from practice.

1. Dalibor Vesely, foreword to catalogue *Material Imagination* (Rome: Artemis Edizioni, 2005), p.10. Exhibition of drawings by author held at British School at Rome, 2005
2. John Berger, *Berger on Drawing* (Occasional Press, 2005), p.102
3. John Willett, *Caspar Neher, Brecht's Designer* (London and New York: Methuen, 1986), p.106
4. Yves Bonefoy, 'The Narrow Path Toward the Whole' in *Yale French Studies*, Number 84, Yale University Press, 1993

Opposite
1:100 section of the Magician's Theatre, National Botanical Gardens, Rome, by Sara Shafiei (Saraben Studio). Made from laser-cut watercolour paper, the section illustrates the detailed patterning on the façade of the building, which allows light to filter through the skin and creates a 'glowing' theatre in the hills.

Left
Chapel of St Ignatius, Seattle, Washington, Steven Holl Architects, 1997. Watercolour rendering showing light, colour and transparency of space.

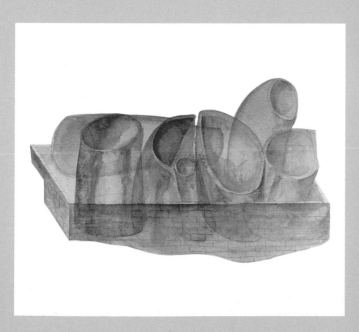

MEDIA

Introduction

This section gives an overview of the range of drawing tools available to the architect, with an emphasis on the representational techniques that may inspire students and professionals alike. The approach taken here is to assume that the computer is only one tool among many others. It explores traditional techniques as well as in-principle guides to CAD software in order to recover the breadth of expression still available to the architect. This is bound not to be exhaustive: it is intended only to cover some key practical tools that can be augmented with reference to other material, printed or online.

In dealing with digital media, the emphasis is to outline principles and approaches to working with certain types of processes and software types. The guides described here are meant to complement, rather than substitute, online tuition and manuals. The most fruitful way to learn technique, however, is through practical exploration, and the following section is intended to inspire a creative discovery of architectural drawing through the practice of drawing itself.

The text is prefaced with comments on drawing surface that affect all drawing techniques. This is followed by an exploration of line in drawings, the most elemental but individual of a drawing's components. When a drawing is developed a little further, the lines may begin to describe form in terms of light and shadow – and eventually render. The second section, render, looks at both manual and digital rendering techniques. Finally, a section on mixed media explores the creative use of combining the two, focusing on techniques that use a variety of materials or processes to create an image.

The characteristics of the drawing surface, its texture, surface durability and colour, are all important elements in the visual qualities of a drawing. This may be true for both manual and digital drawings, depending on output devices. On the whole, manual drawings can take more advantage of different kinds of surfaces: luminosity of the surface is, for instance, particularly important with techniques such as watercolour where thin, translucent coloured glazes allow light to reflect off the paper or gesso surface.

Typically, architects will work on, and certainly print out on, paper. Papers are differentiated first according to the texture and density of their surface. The smoother of these have a surface created by the application of pressured, heated steel surfaces 'hot-pressed' (HP). 'Not' papers (meaning not 'hot pressed', but rather cold pressed – or 'CP') tend to have a coarser (medium or rough) texture.

Both HP and CP papers are also distinguished by weight. As a general rule CP (Not) surfaces are

Below

Landscape Study, detail, using charcoal, pigment and white spirit on canvas.

sympathetic to washes and larger-scale drawings whereas HP surfaces are good for line drawings. Coating either kind of paper with acrylic gesso can make the paper more suitable for other media. Standard 'tracing' paper is best avoided in favour of the translucent layout papers now available. Drawing film is more robust and picks up less dirt. It takes pencil or coloured pencil particularly well and interesting layers can be built into the film by drawing on both sides.

Line

Lines are the most vital components of almost any drawing. Great drawings are read through the character of individual lines and lines come together to define the spatiality of the drawing: lines are like boundaries and as such open up spatial relationships on a page.

The immediacy of a line is the most direct way to visualize thought and observation and as a line drawing evolves, and line weights differentiate, it can express a spatial depth and also define gradations of light and shadow.

Lines are as varied as the instruments used to make marks and the surfaces to be drawn on. Lines can be made with almost anything and media selection depends on individual approach, but as a general rule the combination of drawing surface and drawing tool should be chosen to facilitate a variety of lines; compare for instance the limitation of thin tracing paper with the rich surface of Indian cotton rag paper. For the same reason, you might opt for soft pencil over a fibre-tipped pen, but the final choice will ultimately depend on the nature of the drawing, how detailed it is, its scale and how it is to be seen: will it be viewed close up, from a distance or both?

When drawing by hand, each of us will instinctively make different marks and draw different kinds of lines. These primitive elements of drawing are the most spontaneous reflection of our visual thinking and creative imaginations. They reflect the ways in which we bring together a design as a complex and synthetic process, and in them we can reflect on divergent paths, opportunities and ideas that would otherwise be articulated with difficulty.

Below

Sara Shafiei of Saraben Studio's Anamorphic Tectonics: Magician's Theatre, National Botanical Gardens, Rome. This longitudinal section through the theatre is made using laser-cut watercolour paper with hand drawing and CAD drawing.

TIP PENCILS

Draw by hand with a sharp pencil; reserve pencils softer than 'F' for sketching. Pencil work is a layered process and softer pencils can make the drawing appear too black.

Case Studies: Line

1

1. Here the spontaneity of a line drawing is wonderfully illustrated in the sketch for Open House, Malibu, California (1983/1988–1989), by Wolf D. Prix and Helmut Swiczinsky of Coop Himmelb(l)au. The architects call this drawing an explosive sketch. In a process that recalls the Surrealists' automatic writing from the 1930s, they describe the drawing as having been done with 'eyes closed in intense concentration; the hand act[ing] as a seismograph, recording the feelings that the space will evoke.' The authors go onto explain that 'it was not the details that were important at that moment, but the radiance of light and shadows, brightness and darkness, height and width, whiteness and vaulting, the view and the air.' The differentiated line weights portray a sense of a structure that appears to float, of an ambiguous boundary between interior and exterior and of a spatial sequence that negotiates a steeply inclined landscape.

The sketch is fascinating in its incompleteness; it is both open and closed. It is precise in what it does represent and at the same time open to interpretation and participation by both author and observer in a reflection on the possible worlds that the lines frame in their extensity and depth.

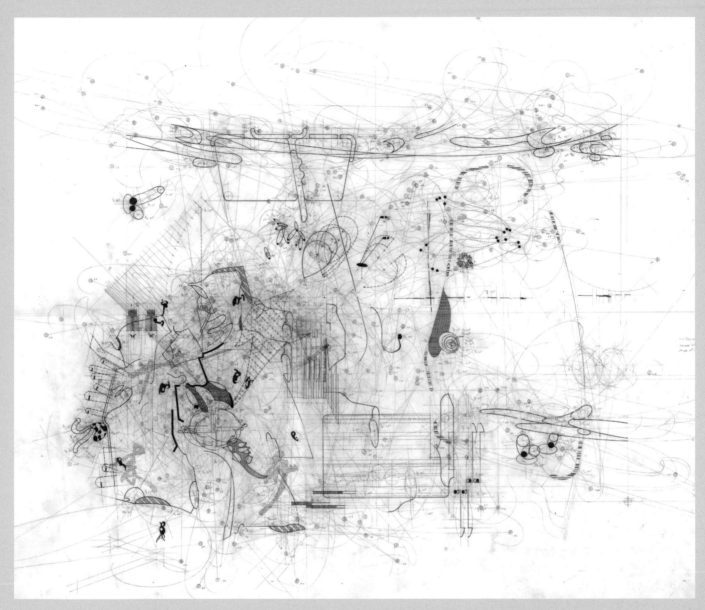

2. Creative line work can be identified in the work of Perry Kulper, an American architect whose body of drawings challenges the way we think about representation. (Several of his works are featured in this book.) Here two line drawings describe a process of thinking as much as a finished proposal. They are done on plastic film (mylar) in a variety of media. Working with specific themes, landscapes and strategies for intervention, Kulper explores the drawing as a tableau that, through line alone, becomes a delicate matrix of spaces that shift in and out of the page; lines that flow or halt and arrest the view. By using each side of the film the drawings emerge as though from construction lines, through lines that describe boundaries; open suggestive patterns of intervention and means of occupation. They are beautiful examples of how, with a limited palette, such a mysterious landscape that is part carefully constructed artifice, and part expressive marks, can be evoked.

The variety of lines in these drawings is in part a graphic tool, and in part a developmental process about the way in which the drawing develops over time. Lines establish the drawing's pace, becoming more or less dense, and take on the qualities of light and shadow. These drawings use lines as tools with which to think about a design; they are open-ended and are vehicles for further reflection that serve a vital role in driving the design forward.

3

3. Lines are an essential part of recording observation and are key components of sketching of all kinds. Observational sketching (and drawing), the rapid recording of a real place, might start out with rudimentary marks, or construction lines that fleet across the page. These should be kept to show how the drawing was made, movements of the hand and the process of observation. Shown here is an observational sketch by Sophie Mitchell. The ink lines are rapidly traced as swiftly as her observation moves across the Arch of Septimus Severus, Rome. Lines are drawn with the handle of a brush and, though they describe nothing more than the broadest of spatial readings, they register wonderfully the process of looking, the moment of expression. They are about the event of observation rather than an illustration of what was seen. In the same way that Kulper's mysterious depths of lines and tone spoke of time and a cycle of drawing and reflection, these action drawings represent a fleeting moment, and capture a speed that has to do with the rapidity of the eye's movements as it takes in the scale, and impression, of the whole.

4

4. Line drawings can become measured through more detailed observation over a longer period. In these kinds of hand drawings lines can become straight cross-hatchings as the drawing develops a spatial depth. This skilful pen line drawing by the architect Kyle Henderson is a fine example of what line drawings of this kind can achieve; retaining a careful balance between the spontaneity of making a sketch and the discipline of careful observation.

Henderson's drawings are lively and inventive but also bear a clear observational resemblance. Like the other line drawings on these pages, it retains an element of incompleteness. Maintaining a balance between areas of detail and other less worked areas of the drawing is an effective strategy.

In each of the drawings on these pages the actual quality of lines plays an important role in how the image is read. An abstracted line implies spatial depth, physical weight and observation.

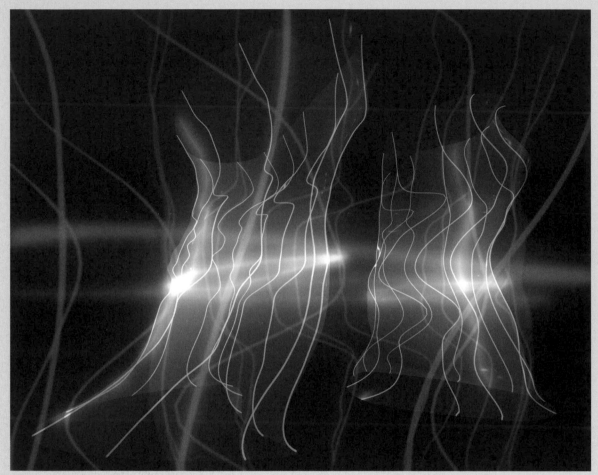

5

5. Sound Travels, Archi-Tectonics. In this study an initial wire mesh model, the undulating surface of which was inspired by a music score, is extended and the forms are rendered digitally using light to investigate form (for another image in this study see page 32).

6. Lebbeus Woods, Berlin Free Zone, 1991.
The image is characterized by expressive line
work and a graphic style that relies on a balance
of line, shadows and light for formal definition.

7. Ben Cowd and Sara Shafiei of Saraben Studio's RAASTA store interior view is a three-dimensional drawing made out of laser-cut watercolour paper, challenging the boundary between drawing and model making. The use of lines is developed three-dimensionally, as a decorative and structural geometry that defines spatial boundaries and interior scale.

8. Eric Owen Moss: Pittard Sullivan, Los Angeles, California.
An effective collage technique that combines photographic and digital renders. Note the restraint of the digital model and the way in which lines of shadow and structure, both drawn and photographed, combine to form an effective collage that is full of movement. The continuity of lines across the drawings makes the collage visually coherent, even though it comprises two quite different drawing techniques.

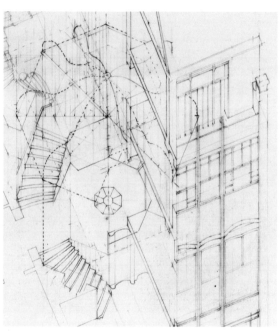

3 Axonometric in pencil and coloured crayon, drawn on both sides of drafting film (mylar). Note the feathering of lines. A feathered line is one where the weight is gradually reduced from thick to thin along its length. Feathering both ends of a line in an architectural drawing gives a line a 'beginning, middle and end'. The line appears to be held at each end in the space of the page, giving the drawing both a sense of precision and lightness of hand. Note also that none of the corners cross.

STEP BY STEP CHARCOAL

Charcoal is a diverse, sensitive drawing medium. This study by artist Helen Murgatroyd shows a charcoal stick used to make a variety of drawn lines, using different pressures and different parts of the charcoal. Textures can be made using a tapping movement or by rubbing the charcoal onto a texture through thin paper. Smudging soft charcoal will give a grey tone, like a wash, which then can be fixed and combined with line work in harder charcoal.

1 Thin charcoal, various pressures

2 Charcoal sideways

3 Snapping as you draw

4 Using finger

5 Thick charcoal, various pressures

6 Soft pressure

7 Hard pressure

8 Repeated tapping

9 Broken crumbs rubbed in

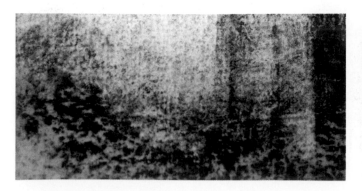

Far left
Soft willow charcoal was used to achieve a variety of marks in this sketch.

Left
Charcoal comes in a variety of sizes and densities.

STEP BY STEP CHARCOAL AND PHOTOSHOP

These images illustrate how Photoshop filters can approximate to charcoal or pastel-like line qualities. All of these effects can be found in the 'Filters' palette under 'Artistic'.

1 'Film Grain' with medium grain and intensity.

2 'Charcoal' with thickness mid range and maximum detail.

3 'Charcoal' with thin thickness, medium detail and lightened. Finally image is inverted.

4 'Conte' with mid foreground, high background, low scaling and minimum relief.

5 'Chalk and Charcoal' with mid charcoal and chalk area and mid stroke pressure.

6 'Chalk and Charcoal' with charcoal area high and chalk area and stroke pressure low.

1

2

3

4

5

6

STEP BY STEP INK

This exploration of marks and lines in Indian ink by the artist
Helen Murgatroyd includes (top left to bottom right): fountain
pen at different speeds and with different sides of the nib, a
blunt metal tool, brushes of different thicknesses and shapes,
a roller, a comb spatula, a pencil end and, finally, the lid of the
ink pot!

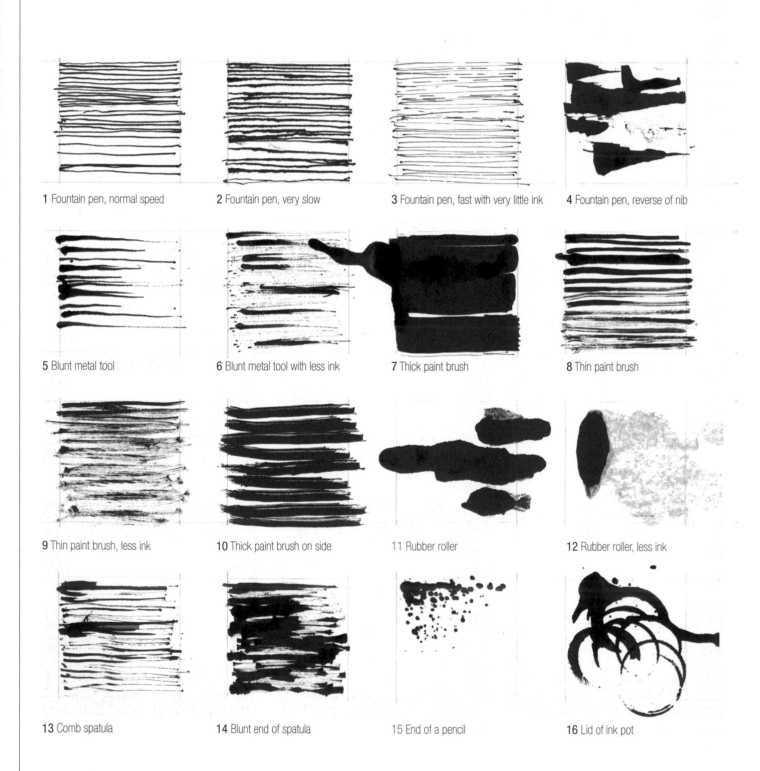

1 Fountain pen, normal speed

2 Fountain pen, very slow

3 Fountain pen, fast with very little ink

4 Fountain pen, reverse of nib

5 Blunt metal tool

6 Blunt metal tool with less ink

7 Thick paint brush

8 Thin paint brush

9 Thin paint brush, less ink

10 Thick paint brush on side

11 Rubber roller

12 Rubber roller, less ink

13 Comb spatula

14 Blunt end of spatula

15 End of a pencil

16 Lid of ink pot

STEP BY STEP INK AND PHOTOSHOP

These images illustrate how Photoshop filters can approximate
to brush-like qualities. All the effects can be found in the 'filters'
palette under 'brush strokes'.

Original image

1 Accented edges

2 Angled strokes

3 Ink outlines

4 Sprayed strokes

5 Splatter

6 Sumi-E

STEP BY STEP MONOPRINTS

Monoprinting is a simple form of printmaking. Basic monoprints, known as direct trace drawings, produce soft-edged lines and tonal effects. Printer's ink is laid onto a surface (for example, a metal etching plate, vinyl, glass or sealed cardboard), paper is placed on top and drawn on, transferring the ink onto the paper as a reversed image.

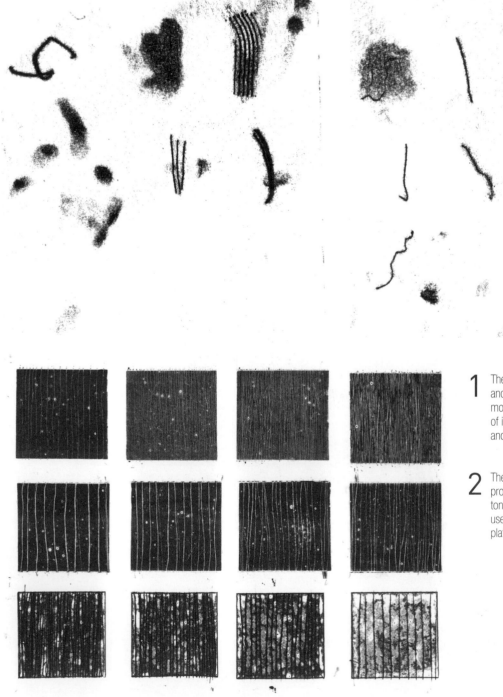

1 These marks and lines illustrate the variety and textural qualities that are possible in monoprinting. Marks are made using a variety of instruments, including pencils, comb spatula and fingers.

2 The simple technique of monoprinting can produce the effect of lines against a textured tonal background. A palette knife or pencil is used to draw onto ink or to take ink off the plate before pressure is applied.

STEP BY STEP PHOTOSHOP

Photoshop is an invaluable tool for visualizing space. In this case CAD modelling software was used to create initial structural forms for a museum and exhibition space. Using transparent layers of texture and colour, the original forms take on material qualities. The 'transform tool' on the 'image' menu is particularly useful in adjusting scale and alignment of overlaid objects and layers.

Left and above
The initial sketch collage was made using Photoshop collage over a form-finding model (above). Using Stylize>Find Edges on the Filter menu, the collage is turned into a simple line drawing (left).

Render

Drawings are the first stages of making. Architectural drawings, as artefacts, evolve to describe light, colour and material surface. 'Rendered' drawings are vital, intermediate stages between the creative imagination and built space. Collections of lines can describe light and shadow; areas of colour, texture and even material fragments can, collage-like, bridge the gap between strategic thinking and material realization. Rendering transforms an abstract drawing; light, texture and colour, both real and fictive, combine to speak of a possible materiality and give a concreteness to the imagined place.

Rendering of this kind is often partial or incomplete. Like a half-finished sketch, the resulting image bears an openness that is as engaging to the viewer as it is integral to the creative design process. This kind of rendering is a natural extension of the line drawing as a process of thinking: exploratory drawings, and to a certain extent sketch models, uncover ways to engage with craft, making and processes of fabrication. Later in the design process rendered drawings can clearly articulate ideas of material and light in order to facilitate detail decisions.

These kinds of rendered drawings are done as the design is in progress. By contrast, a 'final render' has long played an important role in the communication of an architectural proposal. Final renderings are often the most celebrated kinds of architectural drawings and have, through history, used a whole range of possible techniques. Early renderings ranged, for example, from precise pen and ink washes to tempera paintings to

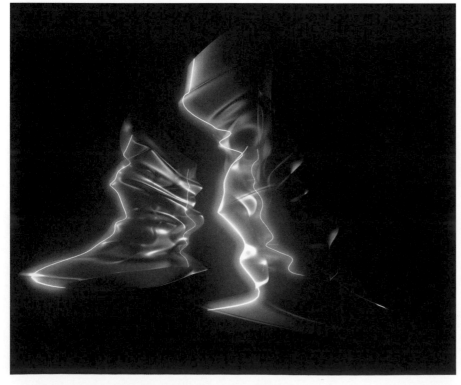

Top left
Peter Sparks' simple pencil and watercolour sketch brilliantly captures the scale, light and materiality of the streetscape. This kind of sketch requires careful adjustment of the amount of water on the page to vary tone between washes and sharp edges.

Left
Sound Travels, Archi-Tectonics. This study shows how effectively form can be described using line, light and shadow alone (for another image in this study see page 20).

frescoes and oil-based paintings. Later, techniques such as watercolour, charcoal and pastel facilitated a more expressive rendering of light, detail and material surface. These images were originally the work of artists and illustrators, but more recently techniques have moved away from such hand-rendered 'artists' impressions', through collage and photographic montage, to computer-generated images (or CGIs).

CGIs vary in character and complexity but this technique is now used for the vast majority of contemporary architectural renderings. More often than not the final image is a made by working in a number of different software packages. Invariably these programs support a formal imagination and are at their best when describing complex forms, structural detail and photorealistic lighting that would otherwise be difficult to represent.

On the one hand, the photorealism of CGI is something relatively new and, using a handful of software packages, the super-realistic render has become a global standard. On the other hand, however, these drawings can often be less than convincing; somewhat formulaic and even unnerving in character. They are not the 'intermediate drawings' that are integral to the creative design process; rather they have a more authoritative character all of their own that represents the building with unerring certainty. Ironically, although graphically almost anything has become possible, there is, at the same time, a level of predictability that means that even the most sophisticated renders can resemble illustrations that lack the engaging capacity of richer drawing forms. A modest idea can appear super real and well-tried visual effects can supplant architectural intention.

Rendering is underpinned by an understanding of chiaroscuro, or how light and dark structure a drawing so as to find and define form, and also to build depth into an eventual colour or tone. In architectural drawing the discipline of 'sciagraphy', or shading in drawings, is the touchstone of many, if not all representational techniques.

On the following pages two works by the artist Anne Desmet, *Poolside Reflection* and *Domus Aurea II 1991*, explore the play of light in space with a particular assuredness. The effectiveness of Anne Desmet's work lies at least partly in her imaginative use of technique and the way in which it connects to the content of the spaces depicted. The purpose of these rendered artefacts is to capture the viewer's imagination; a drawing is there to be explored rather than merely to illustrate; to trigger ideas rather than merely narrate.

The 'presence' of a drawing will in part be a question of content and formal arrangement on the page, but it will also hinge on the way in which the drawing itself is actually made; the material qualities of its surface, its textures and depths. While modern digital techniques tend to reduce surface depth, traditional techniques fundamentally depended on it. Exploiting the properties of natural pigments to have different levels of transparency, media such as tempera, oils and watercolours all work with 'layers' or 'glazes' to create an impression of surface depth. Sometimes almost imperceptible effects – such as the presence of Armenian bole underneath a gilded surface – are part of the way representation has, until recently, captured the imagination of the observer through an investment in surface and light.

Computers present us with a large range of rendering tools and software. These range from basic modelling packages like SketchUp that incorporate an ability to render walls and lighting, to more sophisticated software, like modo, V-Ray or 3ds Max, which is specifically designed to render models efficiently, dealing with complex texture, incident and radiant light.

Photorealistic computer renders are often the result of working across software packages and can be a lengthy process. It can also be useful to develop sketch models digitally that are more quickly 'rendered'. In this sense SketchUp is a popular and useful tool. It is precise as well as being quick to use. Vector drawings from most platforms can be imported and the models can then be exported into additional rendering packages if necessary. Within SketchUp itself are useful guides to sciagraphy, material palettes and components; within Layout, orthogonal drawings can be quickly set up from the sketch model.

Photoshop also remains a vital tool that enables architects to create a vivid impression of a proposal. Photoshop layers can be quickly mapped over views of basic models to effectively represent ideas and take designs forward. The featured interior of 'Revolution Manchester' for instance (see page 31), was rapidly put together in Photoshop as a rudimentary collage over a basic model done in Rhino. This preliminary drawing initiated a design discussion, rather than being a final render. The drawing used obvious Photoshop tools that transform and warp material textures, demonstrating that this program, like other digital tools, is equally effective when it is used with restraint.

Case Studies: Render

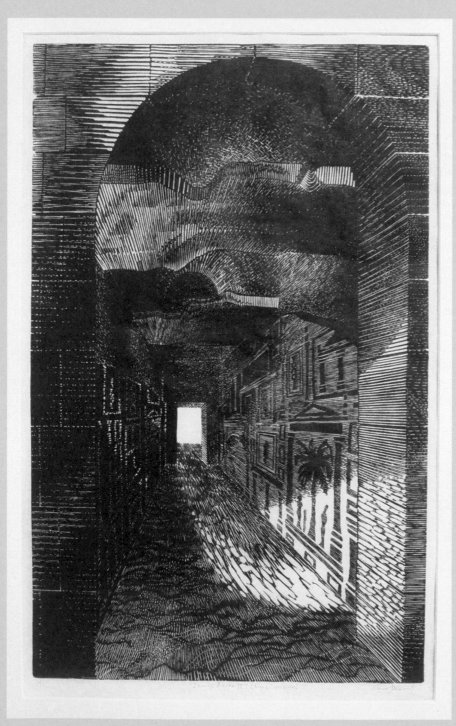

1a

1b

1a. Printmaking is a rich medium for architects to discover the effective use of light and dark. The artist and printmaker Anne Desmet brings a deep understanding of the subject into her architectural works. Here, for instance, is *Domus Aurea II 1991*, a linocut printed in blue/black ink on off-white Japanese Kozu-shi paper. It was developed from tiny pencil and grey wash sketchbook drawings made from memory of the now-underground Golden House of Nero in Rome. It was not intended to be an accurate representation of the interior but more an evocation of some of the light effects, flashes of fresco detail and a sense of the cavernous space, silent abandonment and inky darkness.

2

1b. A second image by Anne Desmet, *Poolside Reflection*, is inspired by the interior of Manchester's Victoria Baths. It is a wood engraving and Chine-collé, printed in black ink. In the cutting of the block, the artist enhanced and exaggerated various light effects observed in the reflected mirror seen in the building and in the photographs. The mirror in question was dented and discoloured, creating a distorted reflection that the artist has exaggerated in her engraving to suggest the effects of reflections in pool water, in former times when the baths were in use. The Chine-collé areas of the print (the buff-coloured paper sections) were added to give a sense of the mirror being a different colour and texture to the wall on which it hangs. The pattern of light and shadow underscores the tonal effect of the Chine-collé to create a sense of the spatial and material conditions of the derelict baths. The abstraction of the mirror-like surface of the drawing engages the imagination and opens up associations that move between light and structure to glazed surfaces and rippling shadows to create an impression of an aqueous world that remains long after the baths have closed.

2. A number of contemporary rendering techniques are best understood as a process of layering. The simplest of these is pencil and coloured crayons. The potential of these techniques is brilliantly demonstrated in the drawings of Eric Parry, one of which, *Elevational studies for an Old Manor House in Wiltshire*, is illustrated above. This sequence of elevations, drawn at a scale of 1:50 in pencil, is delicately balanced between precise, ruled line work, freehand lines and hatching and layered pencil crayon. Together, the simple techniques convey both material and modelling of the building's surface. The drawing is delicate in its execution, but also underpinned with precision, and represents a real sense of the architect's understanding of material, making and landscape.

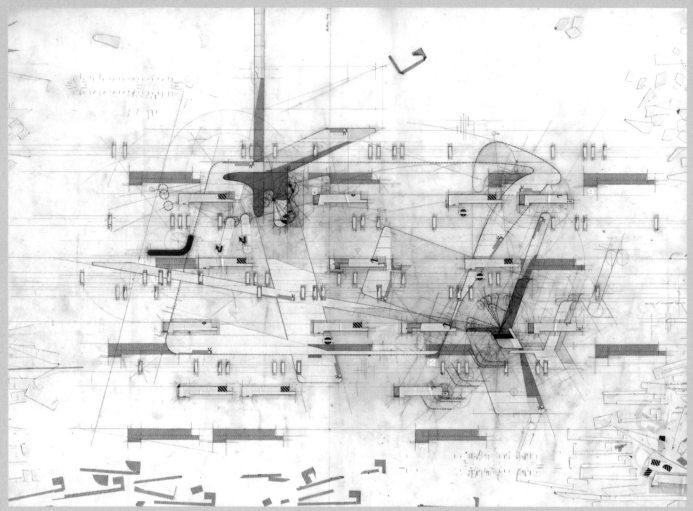

3

3. Perry Kulper develops pencil drawings with a similar refinement to Parry, using line weights of different kinds, and a depth that comes from working on both sides of the drawing film. In this drawing, his line technique is extended to become a more tonal field. The rendering has a flat, graphic quality that contrasts with a more ambiguous reading of overlapping spaces that move across the page. Around the middle of the image the density of the tone increases to establish a space comprised of primary and secondary layers. These are distinguished using different tones. In the background is a light crimson-madder that reads almost like a shadow. In the foreground are more specifically defined shapes, rendered in Naples yellow, and between the two floats a rhythm of grey zones, like elements of structure, made from transfer adhesive tone and occasionally highlighted in white. Finally two crimson-pink elements grow out of the lower shadows and appear to generate an array of other lines and movements. Kulper uses a combination of tonal render and differential line weights to initiate a wonderfully alive spatial dynamic across the page. Relative values of colour and light open and close shapes and movements, like collages of fragments of plan and section to form a composite relief.

4

4. Watercolour, though often associated with smaller observational or illustrative renders, is among the most expressive of techniques and applicable to drawings of all scales. In a watercolour the translucent layers allow the luminosity of the page itself to emerge; light travels through the layers of coloured glazes, is reflected off the page and animates the image. The vibrancy of a watercolour comes from this play of incidental and reflected light within the microscopic depths of its surface. Like ink, watercolour is a challenging technique and depends on a precise control of surface and brush-held water. Here, a rapid sketch by the architects Moore Ruble Yudell represents a plan arrangement. Like most watercolours, the drawing is first mapped out in soft pencil and then liberally coloured, using water in such a way as to encourage the colours to run into each other. Deliberately leaving the paper surface wet in this way gives the impression of both pencil and colour coming together to represent this conceptual arrangement.

5

5. This aerial perspective of the Centre for Music, Art and Design, University of Manitoba, Canada by Patkau Architects is characterized by tonal restraint. The sense of depth in this drawing is in part created by the perspective structure, in part by the selective use of colour and detail in the foreground and then the gradual shift to an out-of-focus image in the background. The understated rendering gives the drawing impact; line and monochromatic drawings can be as powerful as a full-colour photographic render. What is important is that rendering is perceived as a creative and integral component of the process of reflective design thinking, not a simple mechanical application of a software product or technique. Here the level of rendering is particularly well judged, bringing out simple form, façade detail and the broad relationship of the building to the landscape and the overall topography of the city.

6. Lindakirkja, Kópavogur, Iceland, Studio Granda architects. This external render of the church illustrates the architects' intention that in the long, dark winter the perception of the building is inverted, as light from within dissolves the mass of the walls.

7. Delugan Meissel's perspective is interesting in this sense. The drawing is constructed out of a line drawing of an interior. The building has a specific relationship to the hillside and surrounding landscape and this is effectively read in the drawing as it focuses on a montage section of the landscape, collaged in Photoshop as a layer through the glazed wall.

8. Lindakirkja, Studio Granda architects. Internal render showing how the mass of the external shell is evaporated by the light filtering through the vertical slots that make up its walls. Both internal and external renders of this building are effective in their balance between realistic and abstract images.

6

7

8

STEP BY STEP PENCIL CRAYON

Drawings 1 to 6 are of a sketch study of a garden structure in pencil crayon. It was a developmental drawing in the sense that the design was initiated and developed by working through the drawing.

1 An initial sketch study for a garden structure is sketched out by hand using a sharp, F-grade pencil. In this detail you see how the lines are feathered at each corner and how construction lines are left visible. It is important to reduce the amount of erasing, as this compromises the paper's surface quality.

2 Crayon is best built up in layers to create a depth to the eventual colour. It is most effective if shading is done in one direction. The first layer is in Silver Grey and this should lightly cover the entire surface of the drawing with the exception of any areas that are to be left white. The next layer is French Grey that should establish a mid-tone for all surfaces, slightly lighter or darker depending of areas of shadow. Conventionally, the lower the land, the darker the shadow in plan, and water is very dark.

3 Gunmetal and then a blue-grey crayon are used for the areas in shadow. The form and shadow of vegetation is laid in. Earth tones are added to the landscape, which become too dark and are subsequently lightened with a putty rubber. The areas most in shadow are rendered using Burnt Crimson.

4 Golden Brown lightens the ground planes and Burnt Crimson sends others into deeper shadow. The foreground is lightened with Silver Grey and Chinese White.

5 Further depth is built into the landscape, leaving the sketch drawing incomplete and unresolved at this stage.

6 Photoshop then balances ideas that were initiated in the hand drawing in order to finalize this stage of the design with more clarity.

7 The final image remains sketchy — the process is a study rather than an illustration of a final design. This stage introduces a foreground tree (using the magic wand tool to quickly trim and the transform tool to resize). The pen tool is used to draw the shapes that make the foreground shadows.

STEP BY STEP CHARCOAL

These sketch studies are intended to investigate the versatility
of charcoal for exploring a variety of ideas and situations.
In each of the drawings charcoal is used to explore shadows,
establishing form and landscape in terns of chiaroscuro.
A textured surface acts as a key for the soft charcoal.
Darks are laid into the surface and then removed with
a putty rubber in a process that is akin to sculpting clay.

1 This sketch uses the edge of a soft willow charcoal. It is a sketch for a
garden room, enclosed but open to the sky. The sharp contrast in tonality
between the wall and the sky was formed by rubbing some of the charcoal
powder against the edge of a piece of paper. Rubbing out with a putty
rubber forms other sharp contrasts in the same way.

2 The sky is masked and a water layer added using the Photoshop pen tool.

3 This is the finished sketch, developed as a simple collage in Photoshop.
The underlying textures of the initial sketch remain, making the final image
less predictable.

1 This sketch was done using willow charcoal, one of the most sensitive drawing tools, responding to the lightest of touches. It shows an interior looking towards a window opening.

2 In Photoshop, using pen and masking tools, add floor texture and window detail. Develop figure with motion blur to indicate scale.

3 The final image is further developed using lighting effects in such a way as not to lose the material qualities originally implied in the charcoal drawing.

1 This rapid sketch is made with the edge of a harder, compressed charcoal. It is a preparatory study for a garden niche. The combination of this dark charcoal with textured paper is ideal for conveying the kind of material surfaces associated with garden settings.

2 Transparent layers of colour are added in Photoshop as a quick way to explore more detail and a reflective, aqueous floor to the space.

STEP BY STEP WATERCOLOUR INTERIOR SKETCH

The key to a successful watercolour is retaining the luminosity of the paper. Working from the light of the surface, watercolour involves a process of adding translucent washes, working from light to shadow, to create layered colours and depths. In this way the light is always retained in the image. More than anything else, watercolour requires the ability to control the water content in the brush and on the page at every stage.

The following colours are useful for architectural drawings:
• Oxide of Chromium, Cobalt Green, Winsor Green (Yellow Shade), New Gamboge (Gomme Gutte), Yellow Ochre, Naples Yellow, Gold Ochre
• Light Red (Rouge Anglais), Burnt Sienna, Raw Umber (natural), Sepia, Caput Mortuum Violet, Van Dyke Brown,

• Permanent Alizarin Crimson, Cadmium Scarlet, Madder Lake Deep
• Prussian Blue, Delft Blue, Ivory Black, Payne's Grey
Although not strictly necessary, white (or a white gouache for its opacity) is often used to add highlights when finishing a drawing and where the paper has been obscured.

Watercolour has traditionally been the rendering medium for final illustrative drawings. However, it also lends itself to exploratory design, and here are three exploratory sketches using watercolour.

This sequence of three images illustrate the process of making a quick sketch interior study.

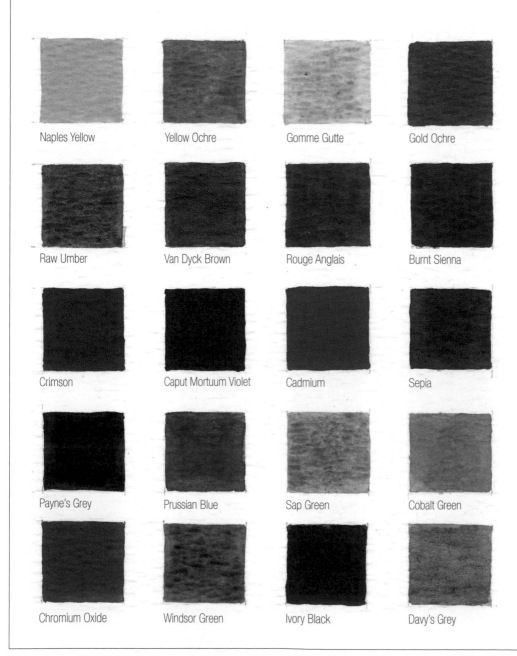

Naples Yellow	Yellow Ochre	Gomme Gutte	Gold Ochre
Raw Umber	Van Dyck Brown	Rouge Anglais	Burnt Sienna
Crimson	Caput Mortuum Violet	Cadmium	Sepia
Payne's Grey	Prussian Blue	Sap Green	Cobalt Green
Chromium Oxide	Windsor Green	Ivory Black	Davy's Grey

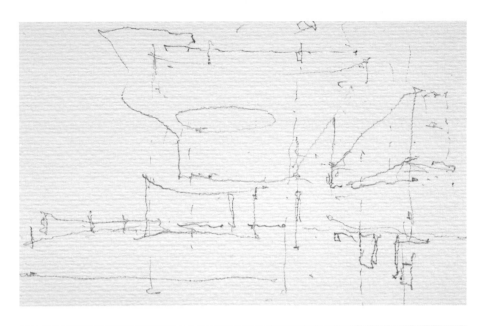

1 First a pencil study roughly maps out what is known about the interior in terms of scale and primary openings.

2 The image is painted over liberally with a light glaze of Raw Umber and Sepia.

3 Detail and reflections are added in Photoshop.

TIP LAYERING COLOUR

Work colours in layers, from lights to darks or vice versa, depending on the technique you are using and the transparency of the medium.

STEP BY STEP PHOTOSHOP
FINISHING A COMPUTER GENERATED IMAGE (CGI)

The following three sequences created by Ian Henderson
demonstrate techniques for manipulating CGI images, creating
shadows and colour-correcting images in Photoshop.

1 'Clay' model render is used to check composition of view, lighting and
model accuracy.

2 The model is textured and rendered in a 3D application. The render is saved
as either a '.tiff' or a '.png' file to preserve transparent areas of the scene.

3 Double clicking on the 'background' layer releases it in order for it to
function like a standard layer. Rename the layer. Drag the sky image into the
working document and place it behind the 'buildings' layer.

4 To lift buildings from the sky create a new layer and position it between the 'buildings' and 'sky' layers. To create the haze, the Gradient tool is used. The gradient is from white to transparent and the layer is given 50% opacity.

5 'Trees are added on a new layer and positioned with move tool. The 'trees' layer is placed behind the 'buildings' layer in the background but above the 'sky' and 'haze' layers.

6 Create two new layers, one for 'tree highlights' and one for 'tree shadows'. Select the trees by control-left-clicking on the 'trees' layer (ctrl-left-clicking on a layer will select everything on that layer). With the selection still active select the 'tree highlights' layer and, using a soft brush, paint the tops of the trees white where the highlights would appear. With the selection still active select the 'tree shadows' layer and, using a soft brush paint the bottoms of the trees black where the shadows would appear.

7 Soft Light layer blend is applied to both the 'tree highlights' and the 'tree shadows' layers with 75% opacity.

8 Grass texture is dragged into the working document using the Move tool (holding down shift while using the Move tool will centralize the imported image to best fit the receiving document). With the 'grass' layer selected create a duplicate layer (ctrl-J). With the Move tool position the 'duplicate' layer adjacent to the original 'grass' layer. Merge down the 'duplicate' layer with the 'grass' layer.

Repeat the duplication of the 'grass' layer until it is at least twice the size of the foreground grass.

Apply perspective to the 'grass' layer to match the grass in the image using the Perspective and Distort functions of the Transform tool (ctrl-T activates the Transform tool. While it is active, right click to access the different transform functions).

Select the foreground grass using the Polygon Lasso tool. (Holding down shift when using the Polygon Lasso tool will constrain the lasso horizontally, vertically and at 45 degrees. While using the Polygon Lasso tool, 'backspace' will undo the last click. The spacebar will permit panning around the image during the Polygon Lasso tool's use.)

With the selection still active and the 'grass' layer selected, create a layer mask to hide unwanted areas of the grass layer (if a selection is made and a layer mask is created it will adopt the selection as the mask. Layer masks allow parts of the layer to be hidden or revealed by painting with white, black or grey. White equates to 100% opaque, black equates to 0% opaque and the shades of grey in between equate to varying levels of opacity).

9 Various layers are created and named to form shadows and highlights on the grass. Black is used for shadows and white is used for the highlights. Each layer is given a different opacity and layer blend depending upon the result required.

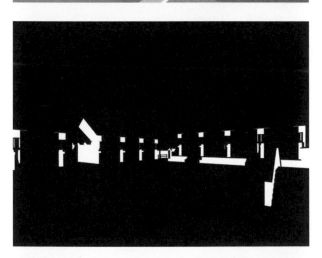

10 Create an 'alpha channel' by rendering a black and white image of the scene from the 3D application. The white isolates a particular material (in this case the stonework) and the black indicates the rest of the model. Drag the 'alpha channel' into the working document and place it at the top of the layer stack. Load the 'alpha channel' as a selection (ctrl-left-click on layer).

On the Channels tab save the selection as 'stone alpha channel'.

11 From the Channels tab load the saved 'stone alpha channel' as a selection. Create a new layer and rename it. Fill the selection with white using the fill tool (ctrl-backspace will fill a selection with the background colour and alt-backspace will fill the selection with the foreground colour).

12 With the new 'white' layer selected apply a Soft Light layer blend and set the opacity to 25% in order to lighten the stonework.

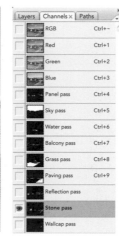

Layers	Channels ×	Paths	
	RGB	Ctrl+~	
	Red	Ctrl+1	
	Green	Ctrl+2	
	Blue	Ctrl+3	
	Panel pass	Ctrl+4	
	Sky pass	Ctrl+5	
	Water pass	Ctrl+6	
	Balcony pass	Ctrl+7	
	Grass pass	Ctrl+8	
	Paving pass	Ctrl+9	
	Reflection pass		
◉	Stone pass		
	Wallcap pass		

13 To change the colour of the stonework select the required colour and create and name a new layer. From the Channels tab load the saved 'stone alpha channel' as a selection. Fill the selection with the required colour. Apply a Darken layer blend and set the opacity to 20%.

14 Load the 'grass alpha channel' from the Channels tab as a selection. Remove the part of the selection that covers the foreground grass leaving only the background grass area selected. Create and name new layers as required to colourize the grass and provide shadows and highlights using the techniques described above.

15 Create a new group in the Layers tab and rename. Drag, drop and position trees as required into the new group within the working document.

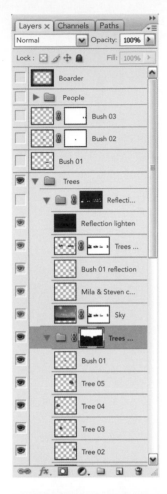

16 From the Channels tab load the 'sky alpha channel' as a selection. With the 'trees group' selected create a layer mask (this will put a layer mask on the group, and therefore the layer mask will affect all the layers within the group). Left click on the layer mask to ensure it is selected and refine the layer mask by painting with black and white to ensure the trees are revealed in the correct areas ('x' will toggle between the foreground and background colours, which helps speed up editing process. To view the layer mask, alt-left-click on the layer mask and repeat alt-left-click to return to view layer. Layer masks can be edited or refined in both 'layer mask view' and 'layer view').

17 Create a new group in the Layers tab and rename. Drag, drop and position elements to be reflected in the windows (ie sky, trees, bushes, people and so on) into the new group within the working document. From the Channels tab load the 'reflection alpha channel' as a selection. From selection create a layer mask on the 'reflection group'. Refine the reflected elements using the techniques described above.

18 Drag, drop and position foreground vegetation. Manipulate vegetation to adhere to perspective and scale using the Transform tool set. Use layer masks as required.

19 Create a new group in the layers tab and rename. Drag, drop and position people. Manipulate people to adhere to perspective and scale using the Transform tool set. Note the direction of the lighting on the people and ensure that it corresponds to the lighting in the scene. Ensure that people sit well in the scene by colour-correcting using image adjustments and/or layer adjustments (levels, curves, hue/saturation, photo filter and so on). Create shadows for people, ensuring that they are cast in the correct direction with regard to the lighting in the scene.

20 To encourage the eye towards the centre of the image create a new layer and rename it. With a large soft brush paint a black border (when using the brush tool, hold down shift to draw a straight line between two points).

21 To finalize the image apply a Soft Light layer blend to the black border and set the opacity to 25%.

STEP BY STEP PHOTOSHOP
CREATING SHADOWS FOR PEOPLE

1

2

3

4

5

1 'Separate figure: ensure the image of the person is on its own separate layer and is named.

2 Create black silhouette: duplicate the 'person' layer and rename it (ctrl-J duplicates a layer). Open the Hue/Saturation dialogue box (ctrl-U). Move sliders for both Saturation and Lightness all the way to the left so both have a value of -100.

3 Distort silhouette: select the 'shadow' layer and activate the transform tool (ctrl-T). Right click over the image and select the Distort function. Grab the upper middle anchor point and move it into position. Double click to apply changes, or hit return. In the layer manager drag the 'shadow' layer so that it lies beneath the 'person' layer.

4 Apply transparency: ensure the 'shadow' layer is selected and set opacity to desired level.

5 Apply Gaussian blur: depending upon the type of shadow required, a Gaussian blur filter can be used to soften the shadow.

STEP BY STEP PHOTOSHOP
COLOUR-CORRECTING A PHOTOGRAPH

TIP LAYER ADJUSTMENTS

Layer adjustments, as opposed to image adjustments, are made because they provide a non-destructive workflow and can be re-edited or turned off at any time. Each new layer adjustment is an additive effect, therefore fine adjustments may be required once all have been applied.

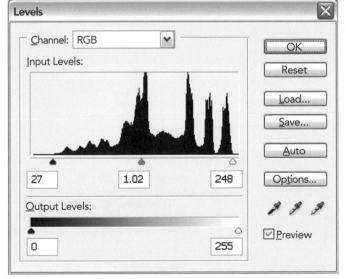

1 The original photo has an orange colour cast and is slightly de-saturated. The Levels layer adjustment allows control over the light, mid and dark tones of the photograph. This can give the photograph more contrast and saturation, and therefore more punch.

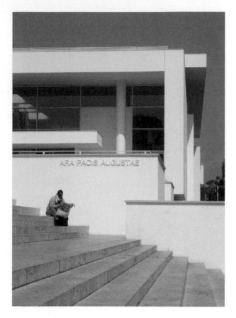

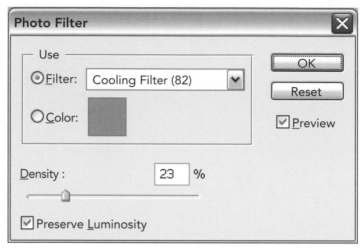

2 The Photo Filter layer adjustment allows control over the colour cast of the photograph. A blue filter is used to counteract the orange cast in the photograph.

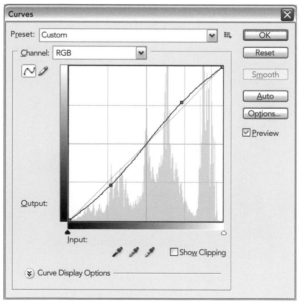

3 The Curves layer adjustment allows control over the image, similar to the Levels layer adjustment. By adding points to the curve, the photograph can be adjusted in many different ways. For this photograph, an 'S' curve is formed by adding and adjusting two points. This brightens the light tones and darkens the dark tones of the photograph and adds more contrast, saturation and punch.

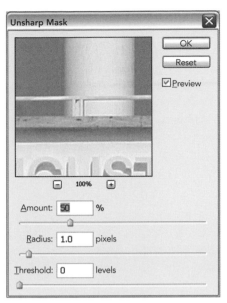

4 Photographs from digital cameras are often very slightly soft and require careful sharpening. The Unsharp Mask Filter or Smart Sharpen Filter are both very good at achieving subtle sharpening. The Unsharp Mask Filter is easy to use and can yield very satisfactory results. The Smart Sharpen Filter is more sophisticated and requires more involvement but can yield excellent results. Any sharpening must be applied with care and a subtle touch.

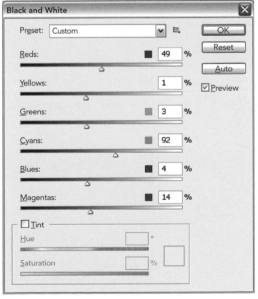

5 To convert your photograph into black and white use the Black and White layer adjustment. The Black and White layer adjustment provides full control over how each key colour group is converted into black and white. Therefore the options are limitless depending upon the effect required.

STEP BY STEP AUTODESK

This short example explores the conceptual modelling functionality in Revit Architecture for creating massing models very rapidly for architectural form finding. The chosen example is a multi-storey office building. This project is intended to be conducted as an instructor-led workshop, therefore only tasks are listed in each section and not the step-by-step instructions to complete the task. Information on tasks can be found in the on-screen help (F1).

To complete the office building in this exercise follow the steps outlined here.

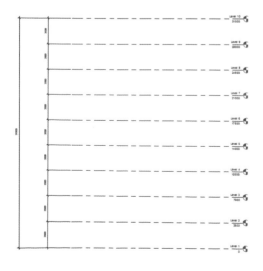

1 Levels:
Open File: DefaultMetric.rte or (step01.rvt).
Create a new project using the New Project button.
Save File: Conceptual Model.rvt.
Define levels in any elevation view: floor to floor heights will be 3500mm.
Change Level 2 elevation to 3500mm.
Create levels 3 to10 with the same 3500mm distance.

Create plan views of new levels: select the View Design bar then select the Floor Plan command and select all levels. Accept default settings.

There is always more than one way to get a good result; try this one: Create new eight times and use the dimension and the EQ to set up the distance of 3500mm.

Save file.

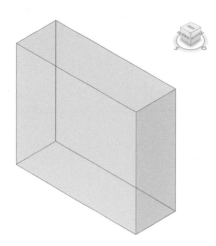

2 Rectangular form:
Open File: Conceptual Model.rvt or (step02.rvt).
Model rectangular form in floor plan view – Level 1.
Massing Design Bar.
Create Mass command.
Accept default name.
Solid Form > Solid Extrusion.

Rectangular Form Dimensions: 36,000mm x 12,000mm x 10 storeys.
Place rectangle in upper half of project file within the elevation markers.

Save file.

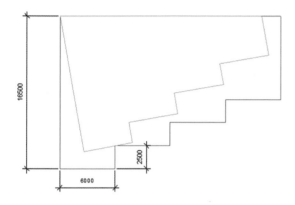

3 Sawtooth form:
Open File: Conceptual Model.rvt or (step03.rvt).
Model sawtooth form in plan view – Level 1.
Massing Design Bar.
Create Mass command.
Solid Form > Solid Blend.

Begin the sketch from lower left corner of rectangle.
Draw Down 16,500mm, Right 6,000mm, Up 2,500mm.
Copy the last two segments three times to create the sawtooth form and drag the last vertical segment to bottom edge of the rectangle.
Close the form with another sketch.

Select the bottom lines.
Resize the sketch lines by 0.9 (numerically); use the upper left corner of sketch as the basc point.
Remove the top horizontal sketch line from the selection set.
Rotate the remaining sketch lines 10 degrees anti-clockwise, using the upper left corner as the pivot point.
Close the form with another sketch.
Set Height of sawtooth form.
South Elevation View.
Select sawtooth form and drag top handle to Level 6.

Save file.

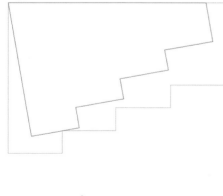

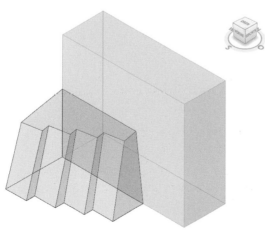

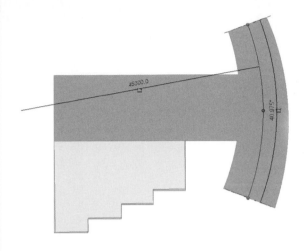

4 Swept blend:
Open File: Conceptual Model.rvt or (step04.rvt).
Edit existing rectangular form.
Floor Plan – Level 1.
Select rectangular form.
Edit button above drawing window.

Add swept blend form.
Add to existing Mass Group.
Select the rectangular form for editing.
Solid Form > Create Path. Pick arc passing through three points mode, path start point above model, path end point below model; Set Radius = 45,000mm (type in).

Create Profile 1: sketch a rectangle (7,600mm wide by 9,000mm) and centre bottom of sketch on path.
Create Profile 2: sketch a rectangle (10,000mm wide by 15,000) and centre bottom of sketch on path.

Join Geometry (rectangular and swept blend).
While still in the Mass Group, Edit mode (so don't finish now).
Use the Join Geometry tool to combine the rectangular and swept blend forms together.
Select the rectangular form first; select the swept blend form.

Exit the Mass Group Edit Mode.

Save file.

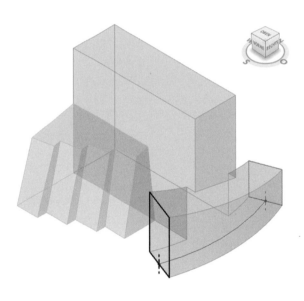

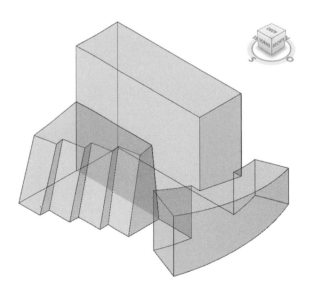

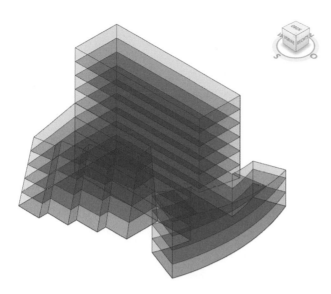

5 Apply mass floors:
Open File: Conceptual Model.rvt or (step05.rvt).
Create floors.
Select massing form.
Click Massing Floor button above drawing window.
Select floors 1 to 10.
Add a tick to the selection set.
Repeat as necessary for each massing form.

Save file.

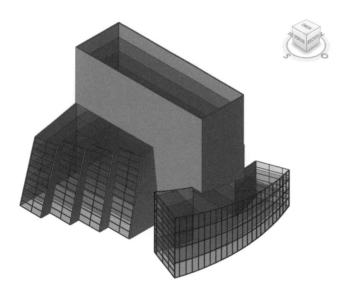

6 Apply walls to masses:
Open File: Conceptual Model.rvt or (step06.rvt).
Create wall by face.
Basic wall – generic – 200mm.
Set Location Line to finish face interior.

Create Curtain System by face.
Curtain System 1,500 x 3,000mm.

Split Walls:
Select the Split tool.
Place the cursor on the vertical edge of the south wall of the rectangular form.
Left click just above the sawtooth form.
Repeat the procedure for the east wall.
Note: You may receive warnings during the Split procedure. Dismiss warnings.

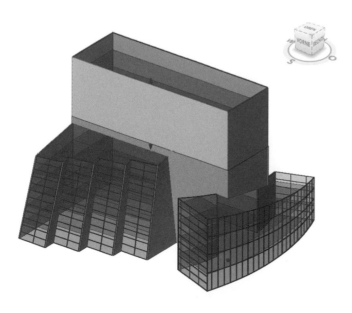

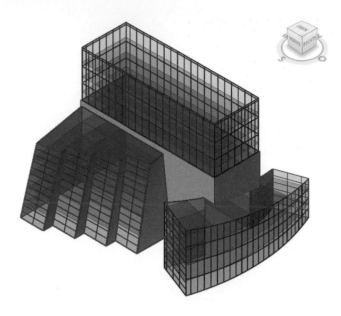

7 Substitute wall types:
Change highlighted walls in red, to:
Curtain Wall – Exterior Glazing.
Note: You may receive warnings during the Split procedure. Dismiss warnings.

Apply 3D Mullions from the Modelling Design bar.
Select Mullion and use the default 50 x 150mm rectangular profile.
Apply to each curtain wall surface.

Save file.

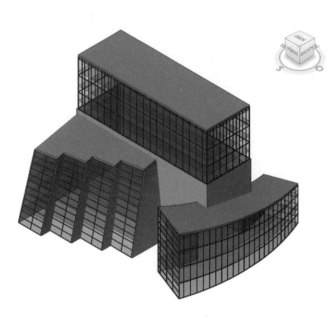

8 Apply roofs to masses:
Open File: Conceptual Model.rvt or (step07.rvt).
Create roof by face.
Sawtooth roof.
Select top of form.
Rectangular and swept blend.
Select tops of forms.

Save file.

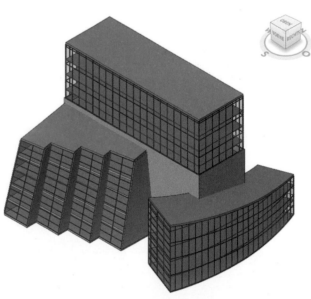

9 Apply floors to masses:
Open File: Conceptual Model.rvt or (step08.rvt).
Create floor by face.
Select concrete – domestic 425mm.
Create a crossing-window selection set of the entire massing model.
Click Create Floors button above drawing window.

Save file.

			Floor Schedule by level						
Area	Cost	Family	Family and Type	Level	Perimeter	Structural	Structural Usage	Type	Volume
306 m2		Floor	Floor slab: Concrete-Dome	Level 1	81000	No	Slab	Concrete-Domestic 425mm	130.05 m3
690 m2		Floor	Floor slab: Concrete-Dome	Level 1	149350	No	Slab	Concrete-Domestic 425mm	293.18 m3
996 m2									423.23 m3
284 m2		Floor	Floor slab: Concrete-Dome	Level 2	78566	No	Slab	Concrete-Domestic 425mm	120.77 m3
690 m2		Floor	Floor slab: Concrete-Dome	Level 2	149350	No	Slab	Concrete-Domestic 425mm	293.18 m3
974 m2									413.95 m3
263 m2		Floor	Floor slab: Concrete-Dome	Level 3	76136	No	Slab	Concrete-Domestic 425mm	111.92 m3
690 m2		Floor	Floor slab: Concrete-Dome	Level 3	149350	No	Slab	Concrete-Domestic 425mm	293.18 m3
953 m2									405.10 m3
243 m2		Floor	Floor slab: Concrete-Dome	Level 4	73864	No	Slab	Concrete-Domestic 425mm	103.48 m3
627 m2		Floor	Floor slab: Concrete-Dome	Level 4	133862	No	Slab	Concrete-Domestic 425mm	266.45 m3
370 m2									369.93 m3
225 m2		Floor	Floor slab: Concrete-Dome	Level 5	71607	No	Slab	Concrete-Domestic 425mm	95.46 m3
485 m2		Floor	Floor slab: Concrete-Dome	Level 5	126367	No	Slab	Concrete-Domestic 425mm	206.02 m3
709 m2									301.48 m3
432 m2		Floor	Floor slab: Concrete-Dome	Level 6	96000	No	Slab	Concrete-Domestic 425mm	
432 m2									
432 m2		Floor	Floor slab: Concrete-Dome	Level 7	96000	No	Slab	Concrete-Domestic 425mm	
432 m2									
432 m2		Floor	Floor slab: Concrete-Dome	Level 8	96000	No	Slab	Concrete-Domestic 425mm	183.60 m3
432 m2									183.60 m3
432 m2		Floor	Floor slab: Concrete-Dome	Level 9	96000	No	Slab	Concrete-Domestic 425mm	183.60 m3
432 m2									183.60 m3
6231 m2									2648.09 m3

10 Schedules:
Open File: Conceptual Model.rvt or (step09.rvt).
Create a schedule.
Wall by type.
Floors by level.
Mullion by length.

Save file.

11 Render model:
Open File: Conceptual Model.rvt or (step10.rvt).
View Control toolbar.
Enable Shadows ON.
Render Design bar.
Select Render dialogue.
Lighting to exterior – sun only.
Set quality level to Draft (1 minute); Medium (10 minutes).
Click Render button to begin the rendering process.

Bonus:
Add site topography object from Site Design bar.
Try new materials.

Save file.

Mixed media

Mixed media drawings use a combination of techniques, mixing hand drawing and different software according to the design process or the intended nature of the final artefact. Accepting the limitations of standard forms of plotting, projected and screen-based media, mixed media drawing offers enormous potential for further developments in architectural visualization.

A hybrid approach to drawing can be particularly useful in the early stages of design. Shifting between hand-drawn and digital techniques can allow a drawing to become a more flexible vehicle for creative thinking, facilitating diverse design approaches in a way that acknowledges drawing as a creative act of discovery, rather than the predictable application of procedures, or illustration; a drawing can reveal something that would otherwise remain undisclosed.

Collages are central to the interpretative drawings that drive design forward. At the same time they are vital tools for understanding context and inhabited space. Collage in its simplest form is made of paper, either cut or torn, and here we illustrate an example of an early study for an urban cinema space and animator's studio (opposite). This drawing began to unravel shifts in scale and a play between real and fictive space that was to be experienced in the urban interior. The torn photocopies were pasted on to brown paper overworked with black pastel and chalks, giving the drawing a more immediate

Below
Saraben Studio's itinerant battery-powered coastal terrain, mixed media, L.A.W.u.N Project #21.

character of being worked when compared to a digitally-generated image. This material quality of the drawing-as-artefact is traditionally the realm of collage.

The collages illustrated on page 65 are early studies for a project, set in London's East End, for a shared external space related to places where cloth was traded, worked and displayed. Taking the context as a whole, design strategies for this marketplace were initiated by making 'material collages' using cloth, thread, hemp, jute and dressmaking patterns. These kinds of drawings were 'registers' of a thought process, at a stage where the final proposal was unclear, as much as they were drawings that expressed the material and fragmented spatial conditions of the context.

Other collages, related to later stages of this project, are made entirely of torn tissue paper. These sketches are scanned and developed in Photoshop to show a more controlled structure, areas of light and articulate spatial proposals. The space and colour of these 'paper walls' preceded the detail development of the project.

The architects Buschow Henley explore the process of photographic etching using black and white CAD images (see page 108). This kind of process is interesting in the sense that it allows for precise drawings to be reproduced that have an attractive surface quality, either through the paper itself or through the layering of printing inks.

Below

The simplest and most immediate form of mixed media drawing is collage. Here is a collage done at the early stage of design development, using torn paper and charcoal. Wrapping paper creates a useful brown background mid tone.

TIP DIGITAL IMAGE SIZE

When working on hand and digital drawing techniques save time by making sure the image size is no bigger than that necessary for the output device settings.

Case Studies: Mixed Media

1

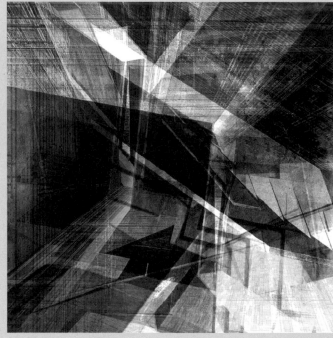

2

1. The work of the artist Anne Desmet shows how printing can be used with the expressive freedom of collage. *Interior Shards*, for example, is a wood-engraved print and gold leaf collaged on to ceramic tile. It is made from small cut sections of other engravings combined to create a sense of entering one of the pool areas of the derelict Victoria Baths in Manchester. The paper pieces are collaged on to a rust-red ceramic tile to generate a sense of the colour of the brickwork of the interior (sandstone); the fact that the interior is also extensively tiled; and, at the same time, to suggest a sense of the exterior brickwork which has distinctive, decorative, brick-red and white stripes. The fragmentary nature of the collage is also intended to convey a sense of the disrepair of the building that it depicts.

2. In contrast to craft-based technique, textures and model photographs can be used effectively to create convincing fictive space in digital mixed-media drawings. This image by the architect Janek Ozmin works with model photographs and material textures to develop extraordinary digital collages that represent spaces that are part found in the scale model, part imagined and part revealed through the process of making the image itself. This kind of sequence of a mixed media technique is exemplary in the way that it drives the design process forward.

3a

3a,b. These mixed media drawings, *Interior/ Exterior I* and *II*, formed part of a preliminary study for an architectural project in London. The drawings were made on a timber base that was clad in canvas and layered with gesso. They were initially formed using tissue-paper dress-making patterns that related to what was to be the eventual programme for the project. They were fixed onto the base with rabbit-skin glue. Mid-tones and shadows were formed using diluted bitumen and other materials that related to site studies. Cotton, fabric, tissue and jute gave the drawings a material quality that inspired later stages of the design.

3b

STEP BY STEP MONOPRINTING

The following six sequences by the artist Helen Murgatroyd demonstrate printmaking processes using various techniques and media.

1 Place a small amount of ink on to a flat surface. A piece of glass or Perspex is ideal.

Using a roller, roll the ink out evenly across the surface. Keep rolling until the ink makes a tacky sound.

2 Very gently place a plain sheet of paper on top of the rolled-out ink.

3 Using any tool (a pencil was used in the example shown) draw the image onto the paper. Take care not to place any pressure on areas of the paper where no ink is to appear.

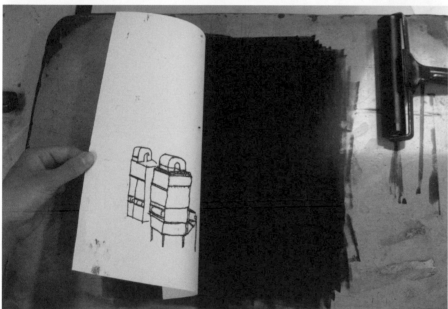

4 Lift the paper from the ink, and your image will be printed in reverse on the other side.

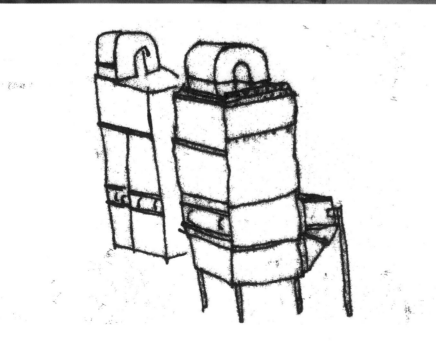

5 Explore different tools for mark-making, such as combs for a series of lines or needles for very fine marks, and try using fingers for covering large areas. Varying the amount of pressure will produce a range of different tones.

STEP BY STEP LINO CUT

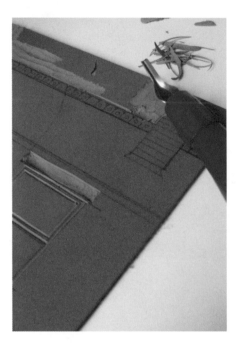

1 Prepare the selected image, working out the areas for each colour. Transfer the drawing onto a piece of lino.

2 Prepare the lino for printing the first colour. Using a lino tool cut into the areas that are to remain 'white' (or the colour of the paper you will print on to).

3 Mix the first colour and roll the ink out onto a flat surface. Only a very tiny amount of ink is required. When you hear a tacky sound when rolling, the ink is ready to roll.

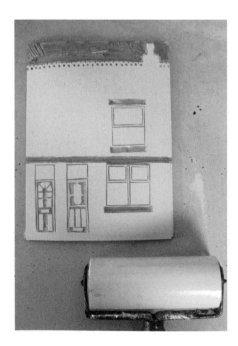

4 Roll the ink very thinly and evenly onto the lino.

5 Place the lino face down onto your paper and roll over the back of it with a clean roller, applying plenty of pressure to make a good, even print. A printing press will give better results.

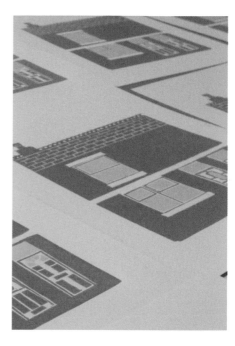

6 Once the lino is clean, cut into it again for the second colour. This time, remove all the areas that you would like to remain printed in the first colour.

7 Repeat the inking up process, with the ink rolled thinly and evenly over the lino as with the first colour.

8 It is important to get the registration of the second print right so the colours line up correctly. If using a press, marking a piece of paper with the position of the lino and the paper is a good way to do it. If using a roller, use the corners of the lino to line it up.

9 Cut into the lino again in preparation for printing the third colour. Follow the same inking up process.

10 When cutting out the areas for the final print there will be hardly any lino left on the tile; only the area for the final, and darkest colour. The tile will be a bit floppy because of this.

11 Repeat the printing process for the final time; the final print is ready.

Stage One

Stage Two

Stage Three

Stage Four

STEP BY STEP PRESS PRINT

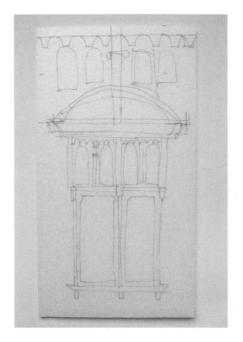

1 Draw the image onto the press print polystyrene tile, making sure not to apply too much pressure.

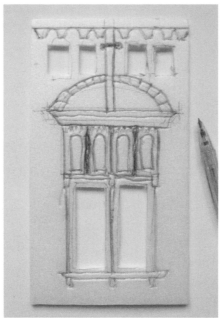

2 Using a ballpoint pen, draw back over the design, making an impression into the polystyrene. Experiment with different tools to make different marks on the tile. A sharp knife can also be used to cut away bits of the tile.

3 Roll the ink out on to a flat surface until you hear a tacky sound.

4 When the ink is making the tacky sound, it's ready to roll on to the tile. Roll it thinly and evenly, making sure you cover the whole tile..

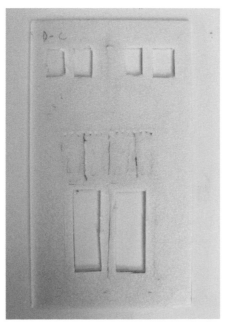

5 Place the tile face-down on to your paper and roll over it with a clean roller.

6 Even the subtle marks made on the polystyrene will be picked up by the print.

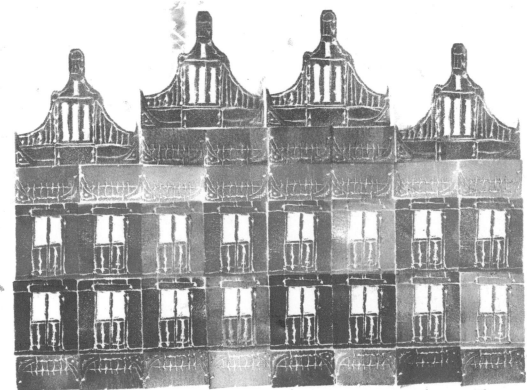

7 Experiment with repeated patterns and different coloured inks. The great thing about press printing is that it is very quick and immediate as the tile takes little time to prepare, but the polystyrene won't last long if it is being repeatedly inked and cleaned.

STEP BY STEP SCREEN-PRINTING WITH PAPER STENCILS

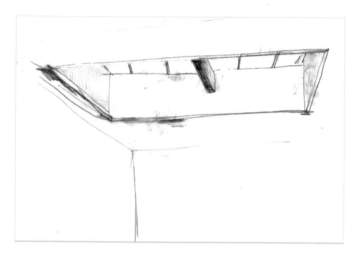

1 Do a plan for the screen-print, working out which areas will be printed in which colour, and how many colours will be needed for the print.

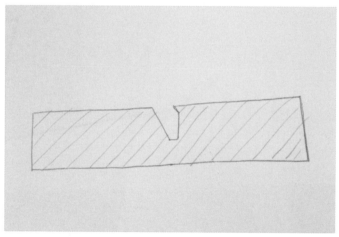

2 Mark out each stencil on a sheet of thin layout paper or newsprint. It is important to use thin paper so the ink will hold the paper on to the screen when printing.

3 Cut out each stencil accurately using a sharp scalpel or scissors.

4 Mix the printing ink. For water-based screen-printing, acrylic paint can be used, which is much easier to clean than oil-based inks. The acrylic is mixed in equal parts with a printing medium to prevent it from drying too fast and blocking the mesh on the screen.

5 Lay the paper for printing on to the screen-printing bed, with the stencil over the top and then the mesh screen on top of that. Ink is dragged through the screen using a squeegee. Print the lightest colour first, working up to the darkest colour; light colours won't print well on top of the dark colours.

6 Stage One

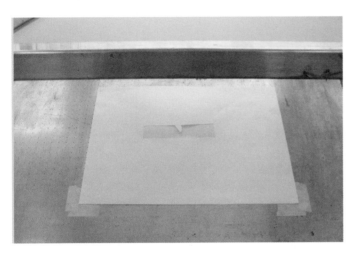

7 Once the first colour is dry, place the print back onto the printing bed, and line up the second stencil, and the mesh screen on top of that. Masking tape can be used in the corners, which is especially useful when printing more than one copy.

8 Stage Two

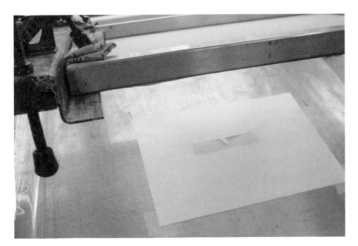

9 Repeat the printing process until each of the colours is printed.

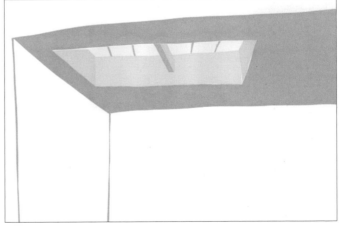

10 The final artwork

STEP BY STEP PHOTOGRAPHIC SCREEN-PRINTING WITH ACETATE SHEETS

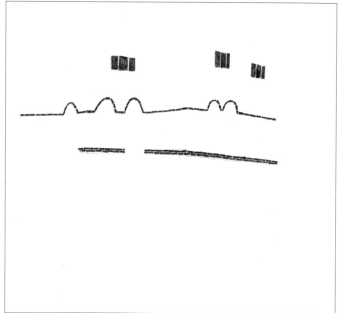

1 Once you've selected the drawing for printing, separate it into different layers, one for each colour. Big contrasts between black and white work best for screen-printing, although a certain amount of tone will be picked up when the image is exposed, so it's worth experimenting with ranges of tone. Photocopy each of the layers onto acetate. Digital files can be printed directly onto acetate using an inkjet printer.

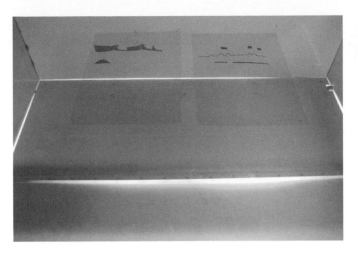

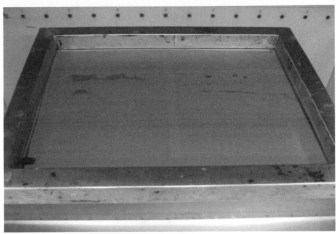

2 Prepare a mesh screen with a coating of photosensitive emulsion and, once this is dry, the image can be exposed on to the screen. Place the acetate sheets on to the exposing box.

3 Place the mesh screen on top of the acetate sheets. In the example shown, two sheets can fit on to the size of the screen that is being used.

4 Clamp the lid of the exposure box shut and expose the screen to the mercury lamp. The light will harden the emulsion on the screen, but the black areas of the acetate (ie the image) will not allow the light through, meaning that the emulsion in those areas will remain soft.

Each different exposure box will have a different strength bulb, so check what the exposure time is on the box you are using before you expose your image.

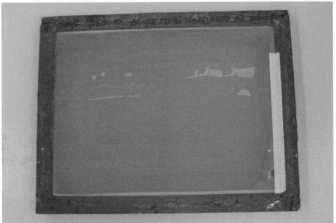

5 Use a power hose to wash away the soft emulsion that has been untouched by the mercury lamp. By washing away the soft areas, a stencil is created on the mesh screen which the ink can then be pushed through to create the printed image.

6 Once dry, the screen is ready to be printed from. The stencil that has been created on the screen is much more permanent than using paper stencils, and can be washed with water several times without deteriorating. A special screen wash is used to remove the stencil once the printing is complete.

7 Clamp the screen securely into the printing bed. Alternatively, you can use a table, but the paper won't be held down as firmly.

8 Position the paper on to the bed. Most beds will have a vacuum beneath to suck the paper on to the surface and keep it exactly in place while you print.

9 Place a reasonable amount of ink on top of the screen, above your image. For the example shown, only half the screen is used, because there are two layers exposed on the screen to be printed separately and in different colours.

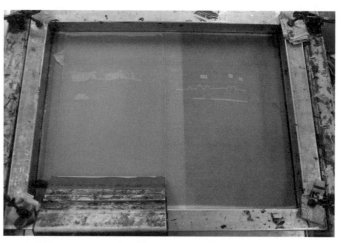

10 Using a squeegee, pull the ink across the screen. On the first pull of the squeegee don't apply too much pressure, just drag the ink until it covers the image, known as 'flooding' the screen. On the second pull, pull in the opposite direction and apply pressure. It is on the second pull that the print is made.

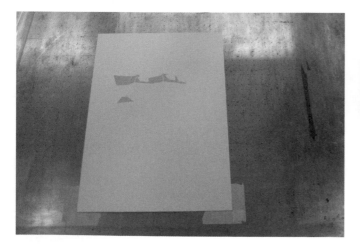

11 Lift the screen, and there should be an even print on your paper. With practice you will find the techniques that give you the best results.

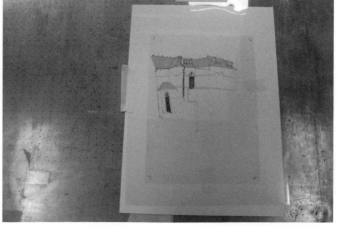

12 Once the first colour is dry, a second colour can be printed on top of it. Accurate registration can be achieved by first printing onto a piece of acetate taped to the printing bed. Once you have printed on to the acetate, slip the paper (with the first colour already printed) underneath and align the two layers.

Stage One

Stage Two

Stage Three

Stage Four

STEP BY STEP PHOTOGRAPHIC SCREEN-PRINTING WITH DRAFTING FILM

1 Instead of transferring the image onto acetate the image could be drawn directly onto drafting film, a transparent film with a texture that you can draw onto in many different mediums. In this example, charcoal is used. Prepare the drawings on different sheets, as with the acetate example.

2 Place the drafting film and screen on the exposing box and expose the screen in the same way as for the acetate. Because the drafting film is very slightly opaque it will need to be exposed for longer than the acetate.

3 Once the screen is prepared and dried, the same printing process is followed as for the other methods of screen-printing. Many of the marks and tones are picked up using this technique. It's worth experimenting with the exposure times to get the best possible print.

STEP BY STEP MODEL/COMPUTER COLLAGE

In the following sequence architect Janek Ozmin skilfully brings together model images and related textures that are combined in Photoshop. The collage is a spatially rich study that records a process of thinking and establishes a broad spatial topography, inspiring further, more detailed studies of the eventual urban and building proposal.

1 Base material: The study started with the generation of a 1:1,000 sketch model for a conference centre made from grey card, beech laminate, black paper generated from a photocopier, photocopied laminate and 1mm clear plastic. Ozmin took a series of digital photographs as part of the exercise with the intention to digitally collage people into the model as part of a spatial study.

2 After completing the first study the original image was moved around using the Rotate Canvas command. After re-orienting, the canvas was stretched using the crop tool in reverse to just off square and the blank space was filled with a stretched section of the brown background.

3 Composite space: the brown section was later masked out in white, which gave the stone floor a reflective quality. Finally a second photograph, taken from a slightly higher position to the composition, was added and positioned 180 degrees against the original image.

4 Keeping the background photograph as a constant, the second photograph was made transparent (above); this immediately increases the spatial experience, creating several areas for the eye to engage with. Focusing on the central portal, the second photograph was copied again and elements of the image were deleted with the airbrush eraser tool (see right).

Leaving the copied layer as transparent increases the visual impact of a portion of the second photograph, giving the overall image a foreground without losing the complexity of the transparencies.

5 Working with the layered composition a space was created using a one-point perspective technique, focusing on the dark section at the centre of the image. Lines existing in the model were extended and three masking planes, two walls and a ground plane were created (above). The masking planes were then made transparent to reveal the space behind (right).

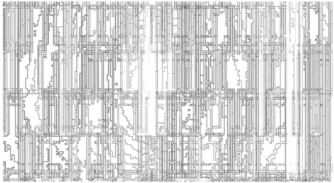

6 Materials: Three materials were selected to be added to the composition. The water and the stone floor were both found using an image search on the Internet. The wall and overhead screen were generated by extracting vector lines from a digital image (below right). Using the Layer Mask tool in Photoshop, the images were applied to the composition and stretched into perspective using the Free Transform tool. Note that the water has been made lighter and colour-saturated compared to the original image (top).

7 Textured space: Reducing the transparency of the masks reveals the background textures. The transparency of the masks will require you to revisit the original materials as with the water, which has now faded and is no longer at a different saturation to the rest of the image.

8 Luminosity: Finally the whole image was selected and copied using the Copy Merged tool and pasted back into the drawing. The composition was adjusted using the Brightness and Contrast tools and the centre of the drawing was saturated using the Burn tool set to 'Shadows'. This removes the milky feel of the previous frame. This image forms the first stage of the process before finishing and printing.

STEP BY STEP PAINTING/PASTEL/MODEL

A sequence of three artefacts – a painting, a pastel plan drawing and a model – represent the design stages for this project.

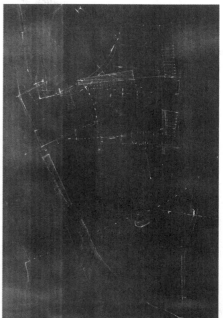

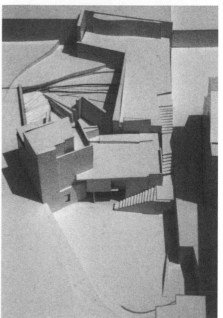

1 The first sketches for the garden and dwelling were translated into a painting made on gessoed plywood. The painting itself comprised earth pigments mixed into glues, waxes and with shell and textures to allude to the garden themes that were to underpin the character of the eventual interiors.

2 A plan drawing to scale (1:200) was drawn from a series of sketches related to the initial painted study. The HP paper was first coated with a soft pastel rubbed into the surface and fixed. Over this was the line drawing in hard pastel drawn partly by hand and partly with ruled lines, with the paper held vertically.

3 Final image taken from study model showing vertical sequence of garden arrangement.

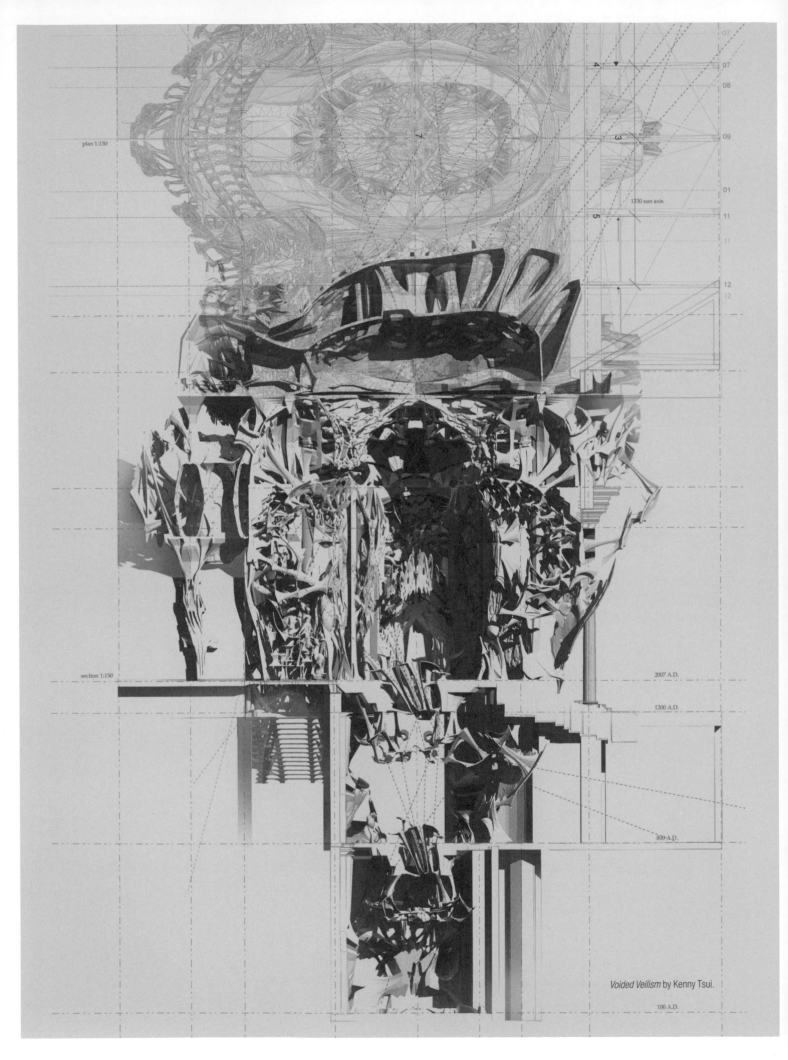

plan 1:150

1330 sun axis

section 1:150

2007 A.D.

1200 A.D.

800 A.D.

Voided Veilism by Kenny Tsui.

100 A.D.

TYPES

Introduction

Architectural drawing combines individual expression and convention in the communication of ideas and information. Part 1 described how drawing techniques could be explored in order to resonate with design approaches, or reflect, in the way they were drawn, the kind of project or idea that was being described. It emphasized the use of mixed media, and integrating hand and digital techniques, as a means of expressing diversity in the design process. Part 2 will explore drawing types, arguing that, in the same way that a project may lend itself to a particular drawing technique, architectural ideas may be best expressed by emphasizing a particular drawing type.

Part 2 covers common architectural drawing types – orthogonal drawings (plans, sections and elevations); parallel projections (axonometric and isometric, dimetric and trimetric projections) and perspectives (one-, two- and three-point). This section also provides an overview of simple digital modelling techniques and an emerging digital drawing type that is made by scripting.

These standard drawing types are preceded with a section on architectural sketches. More than any other drawing type, the sketch remains the touchstone of all our ideas; it is a key tool for observation, reflection and design development.

Below
This rapid pencil sketch shows how lines of different weights can capture a swift impression of a well-known Venetian scene.

The first two sections cover Plans, and Sections and Elevations, and explore the central role that these two-dimensional orthographic projections retain in contemporary design production. These drawings remain fundamental to the process of making architecture: the ability to bring these two-dimensional drawings together in the imagination, understanding their interrelationships and correspondences, remains a creative discipline, vital for the architect.

These conventional drawing types may now appear static, even unimaginative, compared with dynamic, rendered digital models that can visualize complex architectural forms; plan and sectional drawings can be 'sliced' from such models. At the same time, however, these orthogonal drawings can embody a rigour and their strength as tools for thinking through design is that they promote consideration about the content of the programme, rather than its eventual shape, which is privileged by object-based software.

Unlike a finished, rendered model, plans and sections are drawings that depend on other drawings in order to be fully understood, rather than necessarily aesthetic objects in their own right. They may be grouped together with sketches or other three-dimensional artefacts as 'transitional drawings', or artefacts done during a design process.

These transitional drawings are different in character to illustrations; they facilitate a continuity of the creative process, a cycle that is about thinking, drawing and reflecting. By contrast, an illustration is an end in itself.

A designer still needs to be able to translate two-dimensional drawings into three dimensions. At the same time, however, this is now bypassed to a certain extent by modelling software that can simultaneously build three-dimensional form from other drawing types. Three-dimensional object information is input for each element, allowing the digital model to be created at the same time as plans and sections, which can be subsequently 'sliced through' from the model.

This process opens up new formal possibilities and ostensibly shifts the paradigms of the design process, from the discipline of orthographic projection, and the translation of those abstract drawings in the imagination of the architect, towards more immediate visualization of three-dimensional form. Digital modelling facilitates the formal imagination, where the shape and surface of the building, structure and overall massing can be visualized with relative ease. It is dimensionally consistent, and also allows the immediate visual impression of textures, light and shadows.

While computer-generated drawings have promoted a sense of freedom from the conventions of thinking through more traditional drawing types, they also have their own limitations and the approach here is not to promote one drawing type over another, but to encourage an interchange between traditional and new drawing types – traditional and new ways of developing drawings – in order to open a more creative dialogue between digital and manual drawing.

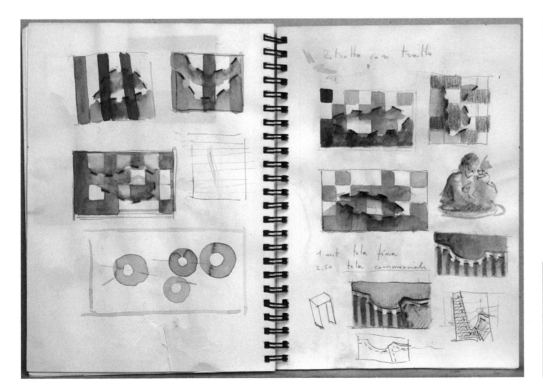

Left
Mario Ricci's sketchbook uses pencil and watercolour to develop an idea that is later the subject of larger paintings and sculpture.

TIP SKETCHBOOKS

A student of architecture should always keep a sketchbook. For students and architects alike, it is a valuable record of process.

Sketches

This section is about an architect's sketches and their uses. Sketches fall broadly into two categories. First are those that simply register the world around us, drawings that are a personal record of visual experience. These drawings may be more or less realistic. Second are sketches that come entirely from the imagination. These drawings are about finding ideas as much as expressing them. They can vary from an irrational doodle to a swift cluster of lines that synthesize a whole project or concept; a drawing that captures the whole sense of a proposal.

Both kinds of sketches are important and while sketching is not an innate ability for all architects, the commitment to visual expression in this way remains a vital skill. The observational sketch is distinct from an illustration: a good observational sketch need not necessarily be illustrative, rather it could be about selection and analysis as much as it is an accurate visual record, complementing photographic analysis in its abstraction from detail. The emphasis of observational sketches is on thinking about the things we see, about spatial and structural organization, scale, light, colour and material. In this sense innate drawing ability is less important than an engaged and focused approach to 'thinking-through-drawing'.

Both kinds of sketches are distinct from, but are often confused with, diagrams. One of the most often referred-to diagrams is the 'concept sketch' that appears at the start of the design process. Like a diagram, this kind of drawing can be reductive in a negative way. The concept sketch is actually a modern invention that derives from the term *concetto* used in Renaissance architecture. But in contrast to the rich combinations of text and image that these ideas once referred to, today's concept sketch tends to be visually diagrammatic and simplistic in content: rather than bringing ideas and images together in a synthetic way (as in the original *concetto*) the 'concept sketch' tends to pull them apart.

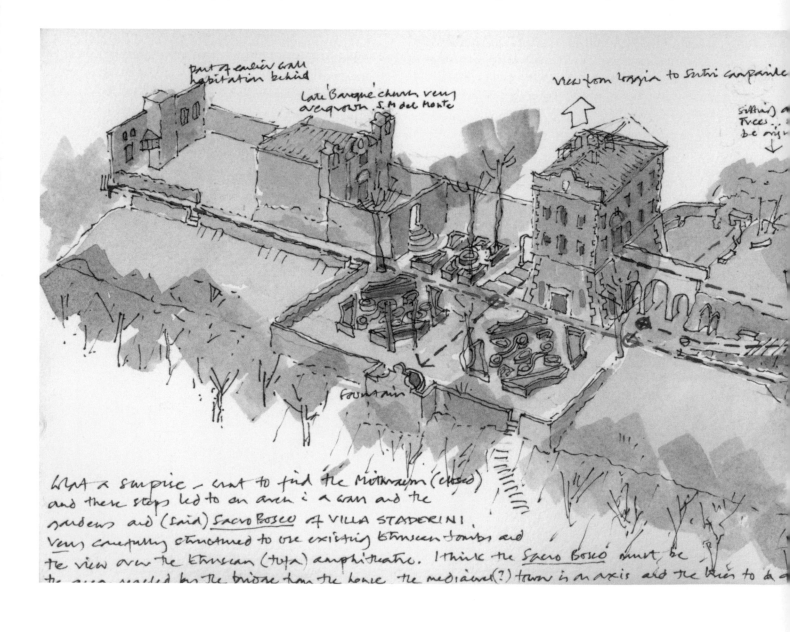

In that sense two common drawing types, the concept sketch and the diagram, are not central to the approach taken in this section. Here we are more focused on the creative potential for architectural drawing and sketches in particular, at the start of the design process, when quick drawings can represent ideas and drive forward the holistic understanding of an architectural proposal. The sketchbook is traditionally the diary of an architect's ideas and still retains a value. Here, illustrations from the artist, Mario Ricci's, and the architect, Peter Sparks', sketchbooks illustrate a narrative of ideas, using pencil, ink and watercolour wash. The sketchbook becomes a domain of thoughts, some of which will develop into further drawings and models.

Below
Architect Peter Sparks' highly skilful watercolour sketch isometric of Villa Staderini, Italy, illustrates how effectively such a drawing can be used to analyze a complex topography.

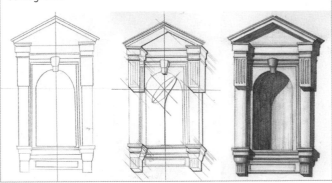

TIP SCIAGRAPHY

Simple projection can form shadows that reveal form. Keep the light source coming from one direction and project through 45 or 60 degrees.

this seems to be a genuine ruin of one town of a castle of some size anyway defending the site

SU QUESTA RUPE GIA SACRA ALLA RELIGIOSA PIETA DEGLI ETRUSCHI E POI AL TRIONFANTE CULTO DELLA MADRE DI DIO SORSE LA ROCCA DEGLI ANGVILLARA DOVE FIORI LA LEGGENDA DI BERTA E DEL PRODE ROLANDO. QUI LA FAMIGLIA MUTI PAPPAZURI COSTRUI LA SUA CASA CHE FU POI DEI SAVORELLI FORLIVESI ED INFINE DEGLI STADERINI ROMANI. INCENDIATA IL V GIUGNO MCMXLIV DALLE TRUPPE GERMANICHE IN RITRATA, TITO STADERINI, FIGLIO DI PERICLE, LA RESTAURO NELL'ANNO MCMXLVI written in the main villa.

two terraced creeks on this lower edge facing Sutri across the Via Cassia

Etruscan Amphitheatre

Case Studies: Sketches

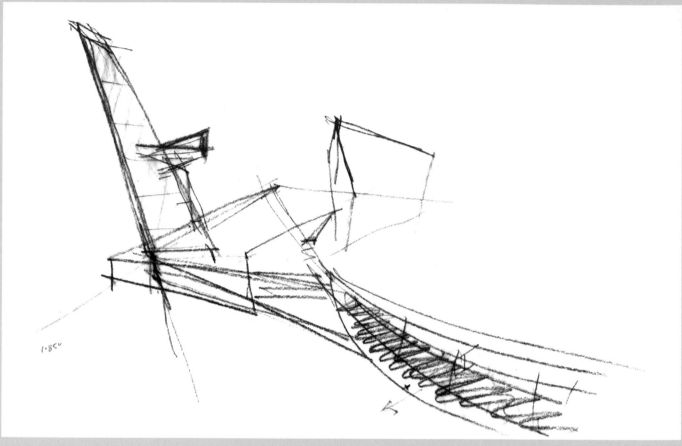

1

TIP ROLE OF THE SKETCH

Sketch in whatever medium is to hand. Emphasize the role the sketch plays rather than the qualities of the sketch itself: a sketch should be used to observe, think or invent.

1. Eric Parry's early pencil sketch for the Southwark Gateway in London exemplifies how, in contrast to a conceptual diagram, an early sketch for a project can embody layers of ideas. In this drawing the first layer is a spatial arrangement: the perspective of a hard landscaped area first captures the sense of a place that may resist the language of street furniture, barriers and traffic chaos typical of the area. The second layer is about more formal articulation, showing an idea for a podium and inclined needle. These are ideas about townscape. Finally the sketch even implies the materiality of the landscape elements (limestone) in the coursing of the needle, and to some extent in the character of the line weights themselves. These layers of thinking come together as a synthetic sketch with a real sense of scale and order that is subsequently worked and reworked, holding together the integrity of the design from urban proposal to detail.

2,3. The perspective sketches by Eduardo Souto de Moura (above right) and Alvaro Siza (below right) are done in ink, using simple lines to capture formal arrangements of buildings that are set against the scale of the surrounding landscape. Like the sketch shown above by Eric Parry, these swift drawings powerfully establish a formal proposal that responds to both programme and context. There is a clear delineation of volumes, entrances and access, and the drawings have a sense of real precision despite the relative freedom of execution where lines overlap and perspective is distorted. The same clarity and discipline is evident in both architects' painstaking attention to detail and the material quality of their built work.

2

3

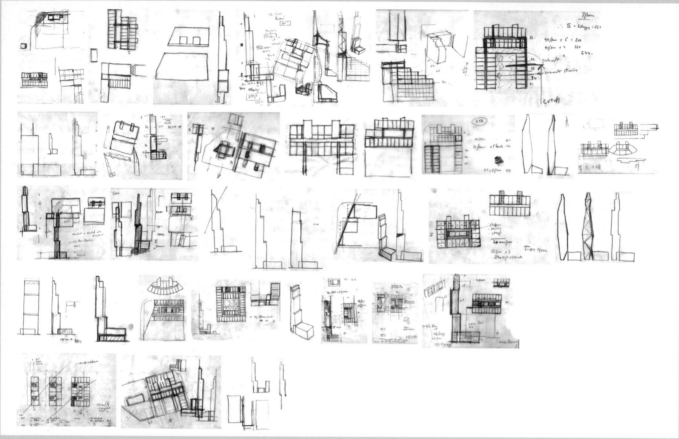

4

4. The previous drawings by Parry and Souto de Moura (pages 92–93) are not stand-alone images as they are presented here; rather they are part of a design development that will include other sketches, models and orthographic drawings. This kind of series is illustrated here in the work of Ian Simpson Architects. The architect Ian Simpson has a remarkable ability to drive even the most commercial of projects with the creative energy that is generated from early sketches. Here is a series of exploratory sketches, traced over a very basic CAD guide drawing showing floor levels and basic building widths. The drawings were targeted at rationalizing and exploring the formal development of the Beetham Hilton tower in Manchester. Each sketch, in pencil on soft detail paper, took about one minute and the whole study was conducted at the drawing board over the course of about one hour. It was later scanned and assembled into a composite sheet for internal review and design development over the subsequent few days.

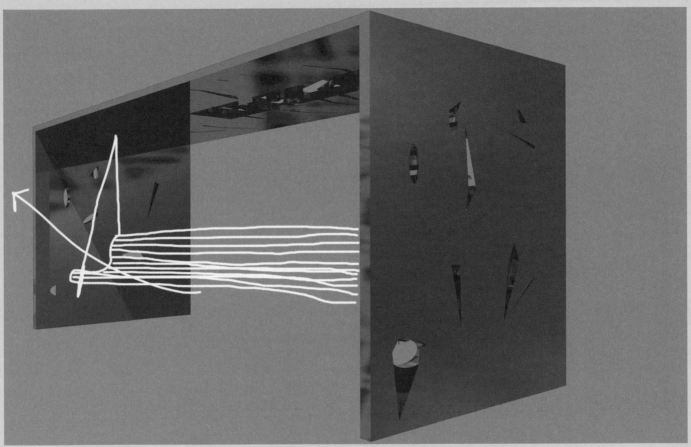

5a

5a, b. In this concept study for garden furniture Will Alsop expresses the desire to think about new kinds of garden furniture altogether. The two images show how a sketchbook study can be combined with a simple digital model. The resulting image takes the initial sketch further in that it represents colour and material qualities and sense of scale. At the same time, in retaining the hand-drawn lines of the swing seat, the computer image retains a sense of openness to the fluidity of design process. This interchange between sketchbook and sketch computer model is fruitful.

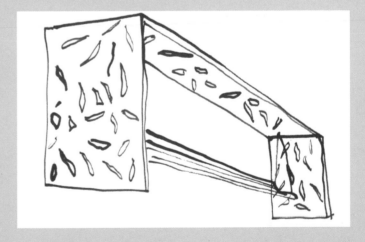

5b

TIP CONVENTIONS FOR PLANS

- Show everything that is seen below the plan cut (typically taken at 1 metre above floor level).
- Differentiate between elements that are cut in the plan with heavier line weights.
- Elements that are not cut, but which are visible, should have a lighter line weight.
- Significant elements that are above the plan cut (i.e. not visible) should be shown dotted.
- Voids to floors below the plan are indicated with crossed lines.
- Voids to floors above are shown with dotted crossed lines.
- Arrows on stairs indicate upwards.
- Like walls, stairs are cut where the plan is taken.
- Show doors in open position.

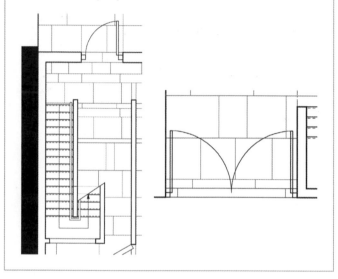

Plans

A plan is a fundamental architectural drawing type. It is a primary organizing device. As a consequence it is the central drawing to a great many projects, and is the drawing through which buildings can most easily be read. At the same time, it is only one part, albeit a significant one, of a whole range of drawings that eventually describe a building in detail.

Technically a plan may be described as an orthographic projection from the position of a horizontal plane. The position of this plane and the scale of the drawing can produce a wide variety of plan types, ranging from landscapes to details. The scale of the plan drawing is important in not only determining the level of detail illustrated, but also the graphic style. For instance, a detail drawing whose purpose is solely to convey information may be appropriately carried out in line, while a plan of a city or landscape may be rendered to indicate topographic information or, in the case of a city, internal and external space. Together the scale and drawing technique of a plan will play an important role in determining what it communicates.

In most projects the plan may be described as something of a 'hinge'; the element about which other drawings turn and the element from which springs a range of spatial studies. If the plan does not work, not only within the frame of the overall intention, but also pragmatically, then other drawings are inevitably compromised.

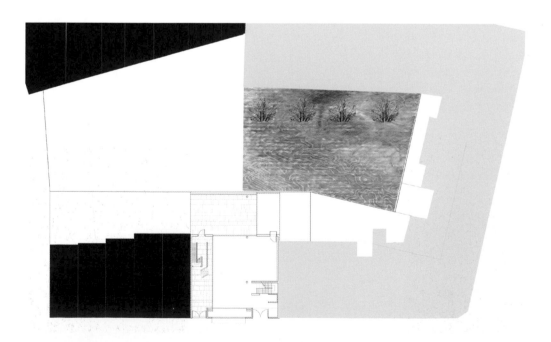

For plans of buildings, the plane of the floor plan is taken a little above, and parallel to the level of, the floor it describes, so that the projected drawing looks straight down on to the floor plane. Of the conventions that govern building plans, the most important concern graphic style and line weights. The key purpose of a plan drawing, in particular, is to communicate information and the key elements of a plan are the vertical planes it cuts through. These can either be shown with a heavier line weight or by rendering the insides of cut walls. All other lines that are visible below the line of the plan are drawn in a lighter line and significant parts of the building that appear above it (a staircase or ceiling plan, for example) are typically drawn as dotted lines.

In contrast to the conventions that have traditionally governed building plans, other kinds of plan drawings can be more exploratory and wide-ranging in their use of technique, evocative in the role they play in generating ideas for the three-dimensional development of the building as a whole. Other kinds of sketch plans are spontaneous and capture the essence of an arrangement.

This kind of drawing is deceptive – though it is quick in its execution, it requires a real understanding of dimension that comes from experience of observing, measuring and drawing existing spaces and objects – measured drawings retain a real value.

> **TIP** SCALE
>
> Above all else a plan should convey information clearly. Always draw a plan to a specific scale (or at full scale in CAD). The conventional scales are 1:10,000; 1:2,500; 1:1,250; 1:500; (landscapes/urban contexts) 1:200; 1:100; (overall buildings) 1:50; 1:20; (individual rooms/details) 1:10 and 1:5 (details).

Opposite
This site plan identifies the proposal by blocking out the context (on the left). Alternatively, one can choose to render just the external spaces (as on the right of the plan), but not both internal and external spaces.

Below
Line weights differentiate elements in the plan, such as walls. These can either be shown as heavier lines (as used on the left side of the plan) or solid filled lines (as used on the right side of the plan).

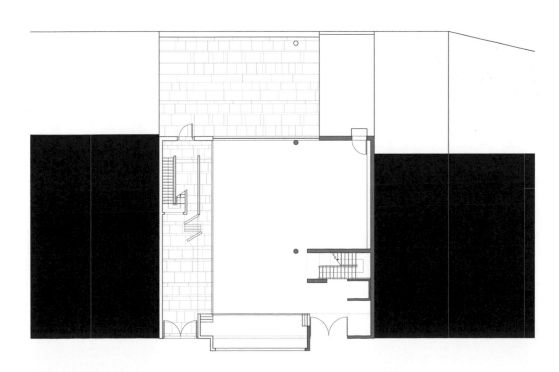

Case Studies: Plans

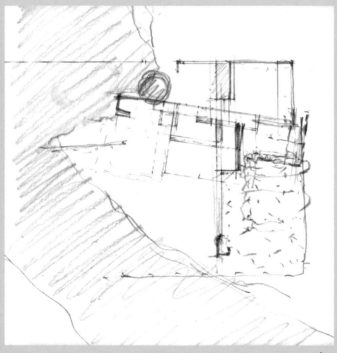

1a

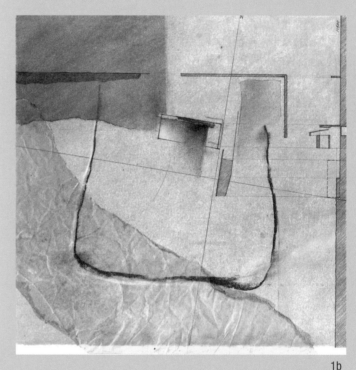

1b

1a. The idea of a plan as a creative device is no better illustrated than in Philip Meadowcroft's sketch plans for a residence. Here two plans are illustrated as a sequence. The first is an expressive drawing on tissue paper in soft pencil. It shows how the architect's first ideas of a spatial arrangement are translated into their preliminary visual form. There is a wonderful range of line quality between sharply ruled lines through to construction lines, fine lines that represent openings or transparent materials, to firm marks that indicate mass or structure. Into this plan can be read layers of thinking that range in scale from the articulation of the hearth to its location in a landscape. The differential line weights annotate, with a variety of intensity, a complex three-dimensional thinking within the conventions of an abstract plan. What is particularly skilful here is that the expressive drawing retains a precise sense of scale (1:500) and this is key to the reading of the overall drawing.

1b. The precision is further developed in the subsequent plan drawing on coloured layout paper. Here, the drawing takes on a material quality, to reflect the importance of the landscape to the overall strategy. From a mid-tone plane of pale yellow, the drawing uses colour and texture (paper creases and pastel) to differentiate tonal background and imply topographic depths. Walls are hatched and ruled as articulation of detail becomes more definite and at the same time the drawing, in the contrast between the precision of the hatched structure and freehand lines of pastel and torn paper edges, conveys a sense of the built and the natural.

2. Ben Cowd (Saraben Studio). *Solar Topographies: The Observatory.*
Detailed plan in laser-cut watercolour paper of an observatory on site in Rome at 1:100. The Observatory overlooks the ancient ruins in Rome and focuses on the transcendental experience of timeless space and form. The building acts as an eternal clock and calendar, tracing the path of time through light and shadow throughout the year: steps representing days and months, stones representing minutes and lines attributed to seconds.

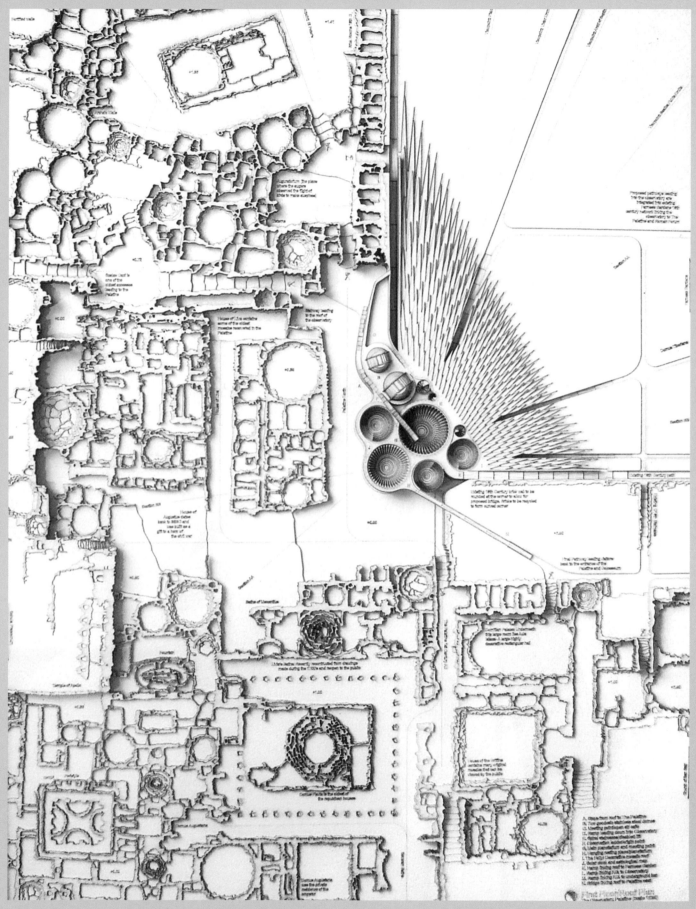

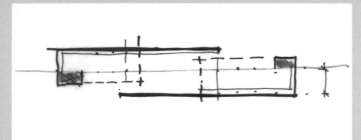

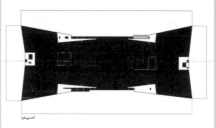

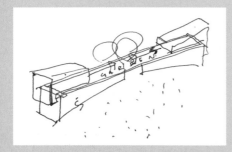

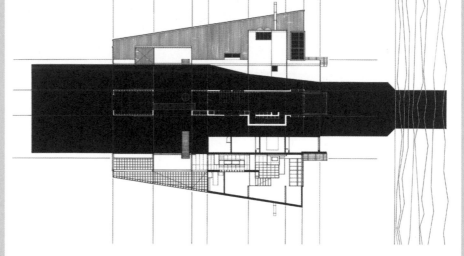

TIP DOTTED LINES

Dotted lines on plans indicate significant elements above the plans view. Crosses on plans indicate voids.

3

3. MacKay Lyons' drawings (top and bottom right) illustrate a different kind of plan, but also the consistent development of similar early sketches (top and bottom left) into final drawings and built form. The architect's plans of the Hill House (top right) and Howard House (above right) are interesting compositionally, combining a simple plan, reversed out (white on black), with elevations and sections. These are engaging graphically and the story of the house unfolds as each section of the drawing, some parts of which are rotated, is investigated. Because of the strong graphics, the drawing attracts attention from a distance and then offers further information, as the intriguing composition invites a detailed study. The bold strategy is appropriately monochromatic and in tune with the well-judged formal arrangement of the buildings themselves.

4a, b. The plans for Archi-Tectonics' Gipsy Trail Residence at Croton Reservoir in upstate New York are expressive of an architectural intention. The form of the plan responds to a rocky, lakeside landscape and the idea that at the 'generative core' of the house would be an 'armature' which integrated kitchen, bathrooms, fireplace, heating and cooling systems, and a central music system. The combination of transparent layers and perspective and orthogonal plans then translates ideas of interdependent layers (wall–glass–roof–glass–armature), efficiency of use (armature as infrastructural core) and negotiation of means (interior/exterior, source of sunlight) into a evocative three-dimensional plan that is both informative and suggestive of the landscape, transparency and 'responsive living'.

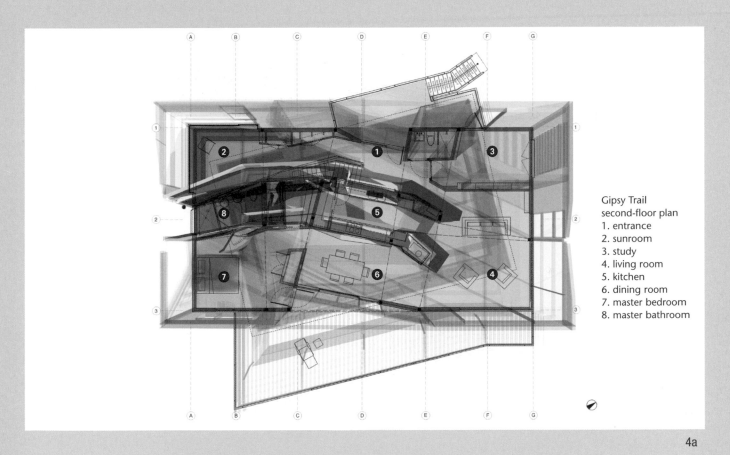

Gipsy Trail
second-floor plan
1. entrance
2. sunroom
3. study
4. living room
5. kitchen
6. dining room
7. master bedroom
8. master bathroom

4a

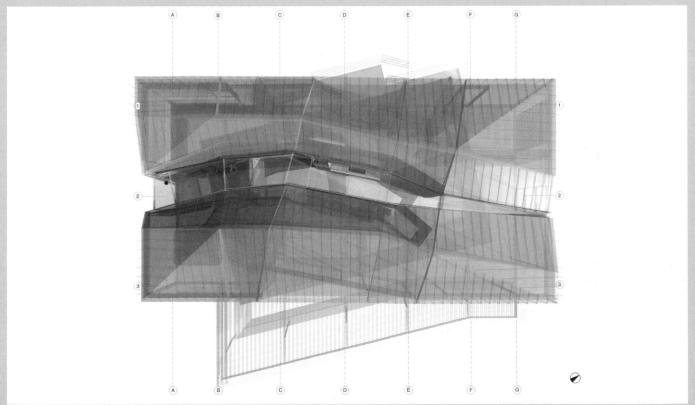

4b

5

5. Zaha Hadid Architects, Kartal-Pendik masterplan, Istanbul (see also pages 119 and 185). The masterplan uses subtle tones and shadows in the render to differentiate between the existing and proposed structures.

6. Sara Shafiei and Ben Cowd (Saraben Studio). These plans, for a market building are partly drawn and partly modelled using laser-cut watercolour paper. The shadows that are created in this way convey a rich impression of the spatial arrangement. The overlaid line drawings clearly communicate other information and the location of this part of the plan with reference to the whole proposal.

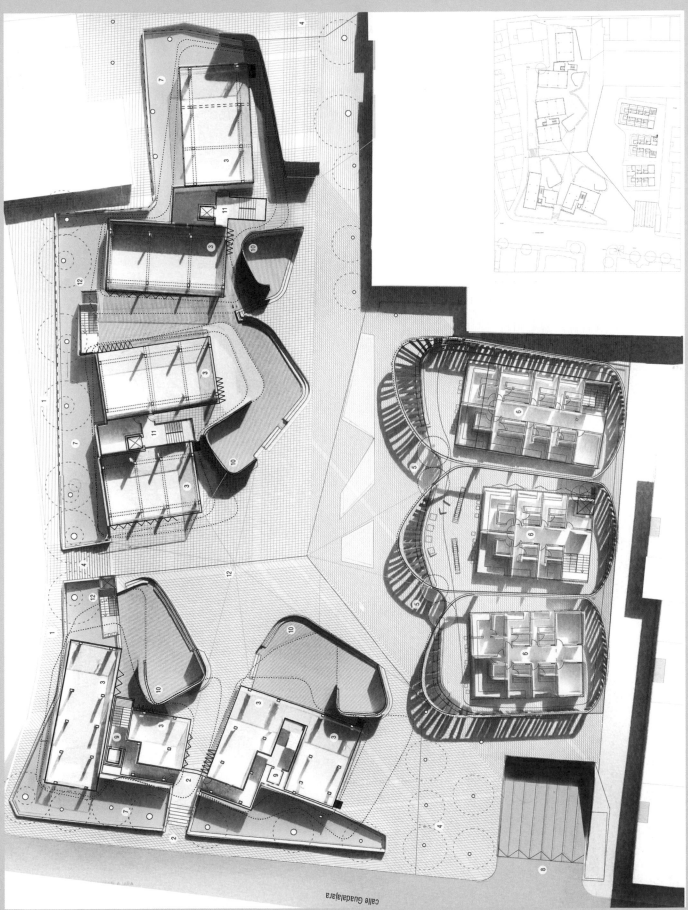

calle Guadalajara

STEP BY STEP SKETCHUP

Illustrations of typical projections that can be taken directly from a CAD model. In this case simple models in SketchUp are taken from 'camera views', isometric, aerial perspective, elevation, orthogonal plans, perspectival plans, orthogonal and perspective sections. A similar function is typically built into all CAD software.

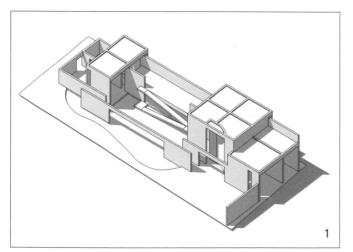

1

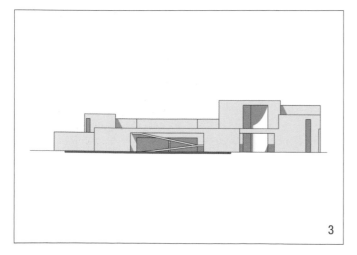

2

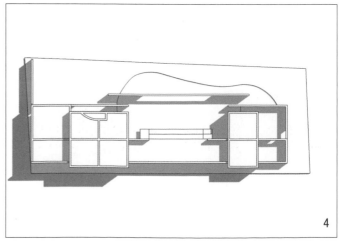

3

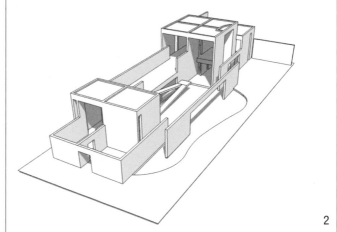

4

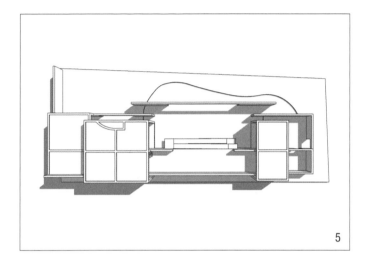

5

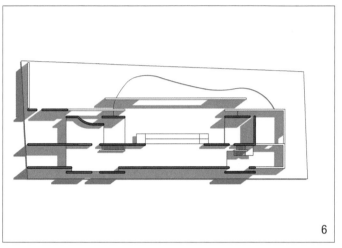

6

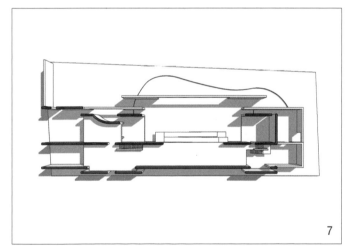

7

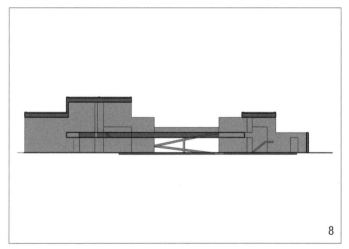

8

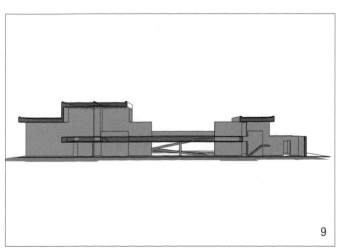

9

All views with shadows turned 'on':
1 Isometric
2 Aerial perspective
3 Elevation
4 Orthogonal plan
5 Perspective plan (upper level)
6 Orthogonal (mid section)
7 Perspective (mid section)
8 Orthogonal section
9 Sectional perspective

Sections and elevations

Sections can be extraordinary drawings. Like plans, they can be the abstract bearers of information, showing heights of rooms, floor thicknesses and constructional details. But a section through a building also embodies a rather special set of conditions: a section presents the 'verticality' of a building: it shows how the building sits with respect to the earth and how it meets the sky. More than any other drawing type, a section declares how the building admits natural light, and describes the thicknesses or transparencies of external walls. Interiors are described in terms of their material surfaces, their depths, passages and transitional spaces. Above all, a section is a 'grounded' drawing: on an uncluttered page a section can be among the most powerful of drawings to describe the organization of space.

Like a plan, a section is an orthographic projection, but from the position of a vertical plane. Sections can be cut anywhere through a building, but tend to be taken where there is a significant spatial condition to describe. Internal elevations will appear between floor plates and an external elevation appears in a section where the vertical plane is taken from outside, or partially outside, of the building.

A section generally offers itself to more elaborate rendering than a plan, partly because of the elevations that are described, but also because a section is a key

Below
Alsop Architects' striking elevational renders are full of colour. Sciagraphy helps describe the depth of façade, and the graphic pattern is emphasized by the simple rendering of ground plane and background.

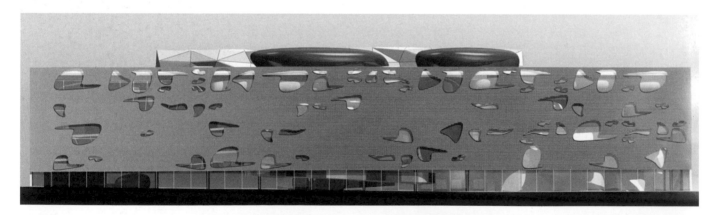

drawing to describe the way in which light works in a building. In rendering light conditions and material textures internally or externally, the section can convey a concrete sense of the proposal.

How contextual a section is depends on the scale of the drawing. At a large scale a section can describe a building's relationship to a whole landscape or a city. A landscape section highlights changes in ground level but is otherwise not the best drawing to describe garden or landscape context, as in a section it will only appear in elevation. On the other hand, an urban section can be an important key drawing, since it will show the scale and proportion of internal and external spaces, and how public and private space is mediated by a vertical layering of the building.

Detail sections are fundamental to conveying construction and technical performance; these kinds of drawings are about clarity of information and are drawn with precisely calibrated line widths.

TIP PLACING A SECTION

A section is best located at the bottom of a sheet. Don't draw anything above it; it should represent a transition between ground and sky.

TIP SECTIONAL CUT

Key to a section is the definition of sectional cut. Take the section through significant spaces, not through structure or secondary circulation spaces.

Below
Eric Parry's sectional elevation is done in pencil, pen and ink wash. The ink skillfully depicts landscape, light and shade. Sciagraphy helps define façade depth and form.

Case Studies: Sections and elevations

1

1. The first sectional drawings featured above are by the London-based architect Philip Meadowcroft. Both are rapidly drawn sketches, direct expressions of thought translated into fleeting lines on paper. In the first pencil drawing it is remarkable how such a minimal group of lines so powerfully describe a sectional strategy. It portrays a stepped landscape, with a building form cut into it. A garden, framed by a dark wall (possibly a hedge), addresses a walled garden. There is little more than an essential set of sectional ideas that describe the verticality of a garden room, and the light and materiality of a landscape setting.

2. Architects Buschow Henley have developed a photographic etching technique for printing CAD images as etchings using an aquatint process. The images produced have a beautiful, soft quality otherwise unattainable by straightforward plotting.

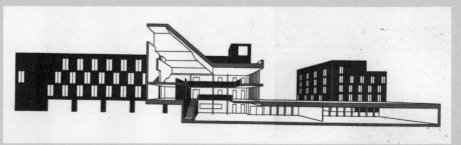

2

3

4

3. Equally sensitive is the elevational section for Chetham's School of Music, Manchester, by the architects Stephenson Bell. These drawings are work-in-progress, but illustrate a stage in a useful design cycle. Production started with the combination of developmental sketches and Microstation drawings in order to develop the three-dimensional massing and proportions. The drawings were then imported into Photoshop where they were tested in terms of materiality and light and shade.

4. A section elevation by Hodder Associates of their proposal for St. Catherine's College, Oxford. It successfully represents the building in its context by combining a simple line section with Photoshop collage. The effectiveness of such drawings depends on the use of filters and masks to tone down photographic context.

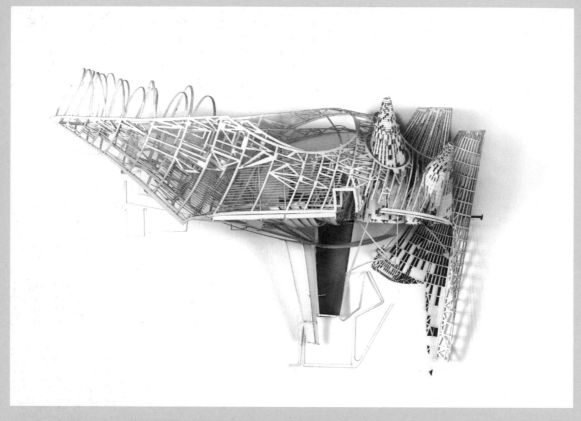

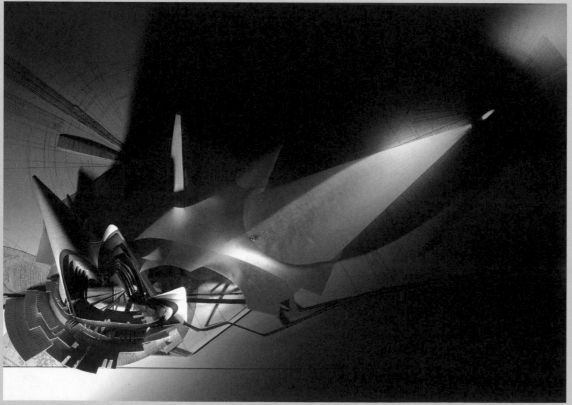

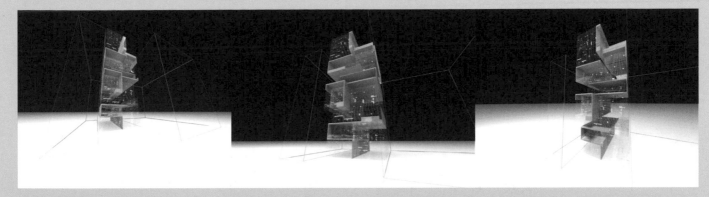

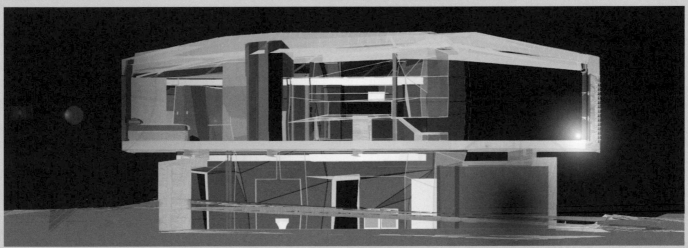

6

5. Sara Shafiei and Ben Cowd (Saraben Studio). Magician's Theatre, National Botanical Gardens, Rome. Elevational and sectional studies. Media: tracing paper/hand drawing.

6. The sectional perspectives of Archi-Tectonics' Vestry Street residential infill project, situated in New York's Tribeca district, show how the section can be developed as a key tool to express specific ideas about a project. The form of the project was generated by a response to the varying building heights along the street and was articulated as an overlapping of suspended and cantilevering volumes, here cleverly portrayed through a sequence of elevational perspectives that show how the floor plate, which folds in section, forms the elevation as a pattern of solid and transparent (pixilated) planes.

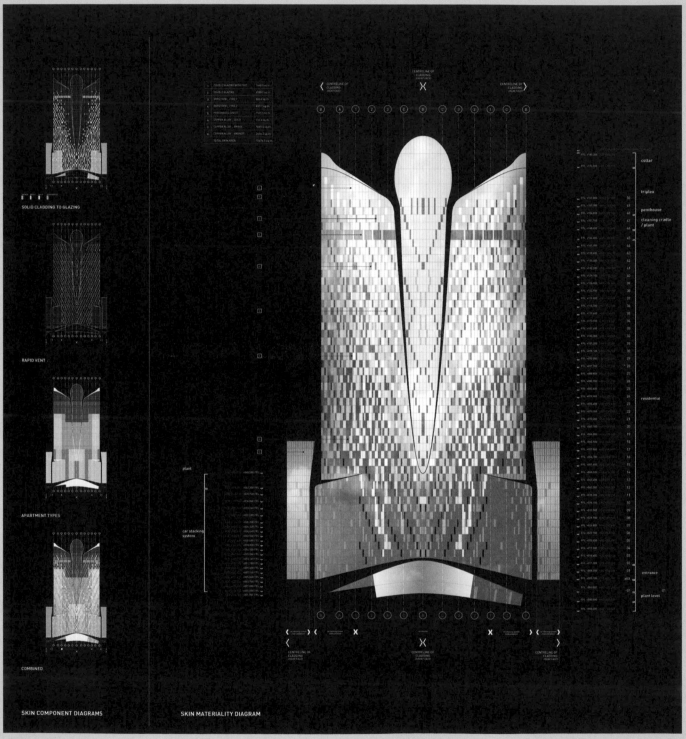

SKIN COMPONENT DIAGRAMS

SOLID CLADDING TO GLAZING

RAPID VENT

APARTMENT TYPES

COMBINED

SKIN MATERIALITY DIAGRAM

7b

7a, b. Ian Simpson's early developmental drawing is equally inventive. It shows an unfolded cladding elevation for a tower project, ovoid in plan. The original elevations drawn in Vectorworks and the 'unfolding' of the façades were calculated and constructed manually: at this stage (2004) Vectorworks lacked the ability to automatically 'unfold' the geometry of the cladding skin. Basic colours were also added in Vectorworks with post-production colour and reflections added in Photoshop.

Axonometric and isometric projections

Axonometric projection is one of the most frequently used drawing types for creating three-dimensional images in architecture. Isometric, dimetric and trimetric projections are types of axonometric drawings and together they form a group of drawing types called parallel projections or paraline drawings. These are three-dimensional drawings, projected using orthographic projections as generators, where all parallel lines in the object remain parallel in the drawing. Objects drawn in parallel projections do not appear to get smaller or larger as they recede, and line lengths remain dimensionally accurate.

Isometric drawings are formed when an object is turned so that all three axes meet the picture plane at the same angle, making the angles between the edges of the building or space 120 degrees. An isometric drawing can be drawn at any scale and so best illustrates the true projected size. In architectural drawings one axis is usually vertical and the other two are therefore at 30 degrees to the horizontal. Most architectural drawings are shown from the point of view of looking down, (though details and ceilings may be best seen looking up) and view a plan that appears to be 'opened up' in the direction of viewing from 90 to 120 degrees. This is advantageous if the drawing reveals an interior, but otherwise

Below
These diagrams were created for the Centre for Music, Art and Design at the University of Manitoba, Winnipeg, by Patkau Architects. Both were modelled in AutoCAD, rendered in VIZ and enhanced in Photoshop. The six large, flexible primary spaces are shown coloured, and the support spaces are grey. To describe the disposition of the primary spaces within the overall building, the building was represented as a transparent cubic form with the coloured primary spaces within it.

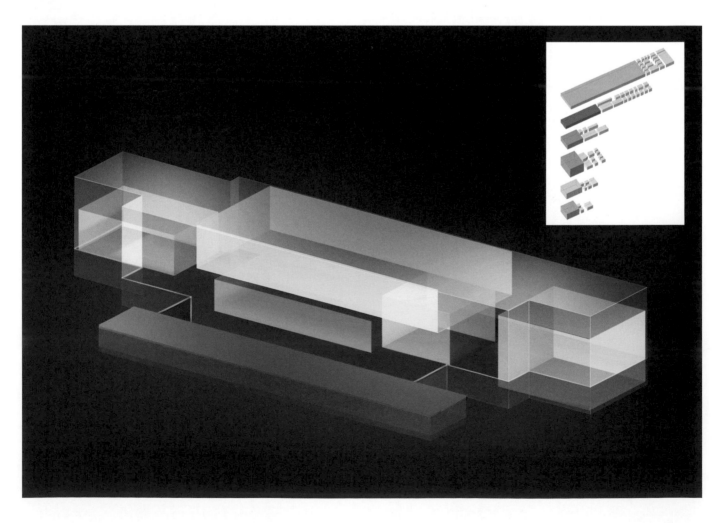

is a relatively rigid projection that requires all three visible planes to be emphasized equally.

Axonometric is the term used in architecture to describe a projection from a plan to scale. Like isometric, axonometric drawing is a form of parallel projection: plan elements are projected vertically and to the same scale as the plan, which is first rotated to provide the intended view. Walls that appear at the front of the plan can be omitted or only partially projected to create a 'cutaway' axonometric which reveals the interior that would otherwise be obscured by uniform projection. The cutaway drawing is a useful tool for describing spatial sequences or narratives that link different levels in a building.

Dimetric projection is rarely used today because it is not a standard projection. For Modernists, on the other hand, the dimetric projection was a common drawing type due to the flexibility of viewing angles it allows. In dimetric projection the image can be adjusted in terms of the scale of axis, and also in terms of two viewing angles, allowing asymmetric adjustment to correct the visual distortion that inevitably accompanies a more rigid projection type. Two of the three angles are equal with the picture plane and the third angle is different, so that two of the three axes appear equally foreshortened and the scale of the third direction

(vertical) is determined separately. Although there is a degree of complexity to drawing dimetric projections, since they require the use of a scale factor and an adjustment of two angles to the picture plane, there is also flexibility otherwise not accommodated in projections of this kind. This drawing type allows an infinite variety of viewing angles relatively simply and so is also popular in computer games.

The least common, but most flexible, of these drawing types is trimetric projection where all three axes appear unequally foreshortened and the scale of each foreshortened side is determined by the angle of viewing. This added foreshortening gives these drawings an unnecessary degree of complexity for most architectural ideas, which can usually be represented through one of the other axonometric drawing types.

Below

Garry Butler (of Butler & Hegarty Architects) has developed a deep understanding of timber joinery through detailed surveys and careful restoration. These on-site axonometric sketches are brilliant examples of how this type of drawing (together with sketch plans and sections) is perhaps the only way to understand complex details of this kind.

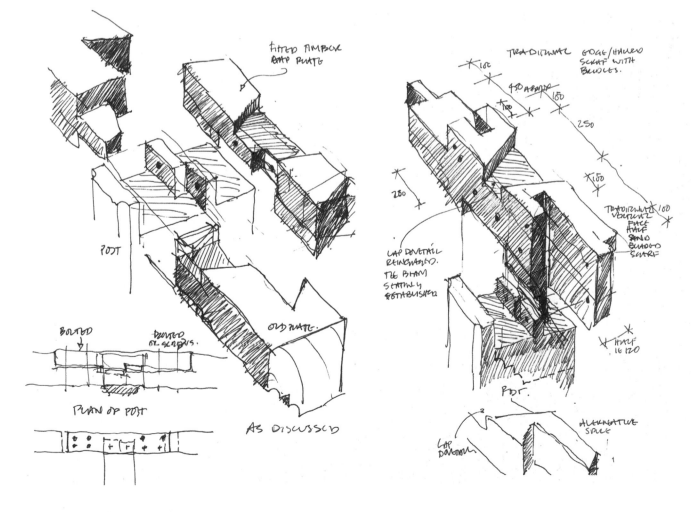

Case Studies: Axonometric projections

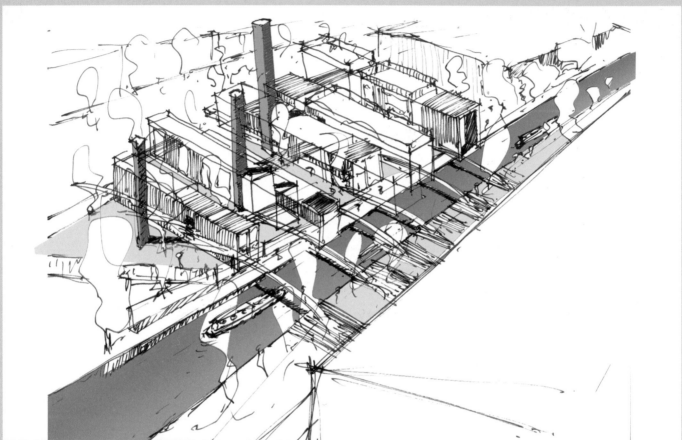

1

1. This sketch isometric study by Kyle Henderson illustrates how useful this kind of projection can be in representing an overall idea in its context. The pen sketch shows not only the form of the buildings, but sufficient detail to distinguish primary material differentiations. The context is mapped out in less detail, but the relationship between the building form and the series of bridges is clearly emphasized, with each bridge casting a well-defined shadow on the waterway below. The sketch is scanned and areas filled in Photoshop represent a further layer to the narrative that is about the landscape, gardens and external spaces interspersed between buildings. The isometric sketch allows this building strategy to be represented as a whole and, because of the 'opening up' of the axis in this kind of projection, offers glimpses into the spaces in between buildings.

2

2. By contrast, this well-known drawing of the
Villa of the Physicist by Eric Parry is a more
developed study using collage, pencil, inks
and pastel. The basic form of this drawing is a
simple isometric. Set against the black external
masonry piers, the mysterious interior seems to
draw its light from the landscape that is sketched
in around it. The contrast in technique here is
important, as the pastel and soft pencil easily
fade into the distance and allow the building
to appear as though integrated within it. In this
drawing there is a wonderful balance between
the constructed and the landscape, between
light and dark and between elements that are
well defined and others that are more open
to interpretation.

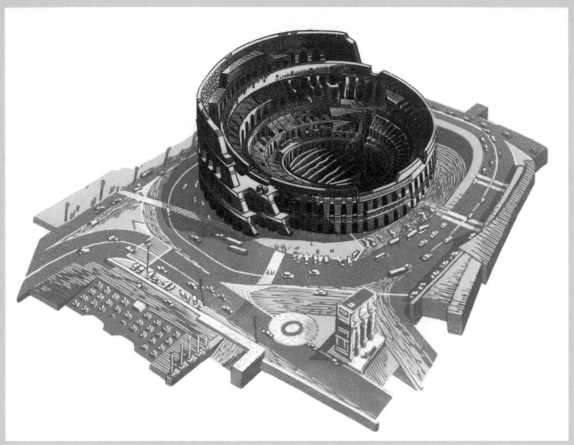

3

3. Inspired by an old postcard, *A Slice of Rome* by the artist Anne Desmet illustrates axonometric projection in an urban situation, in this case to show one of Rome's best-known landmarks in context. The image is made from a lino-cut and collaged wood-engraving on white paper. The road system surrounding the Colosseum is a three-colour reduction lino-cut print; the collaged Colosseum engraving superimposed on top is printed in blue/black ink onto cream Japanese Kozu-shi paper. The surrounding roads are bent, exaggerated or truncated to make a strong overall composition but the image has a wonderful sense of movement that is born out of the detail within the image and the textures and colours of the print.

4. Zaha Hadid Architects, Kartal-Pendik masterplan, Istanbul. These drawings, vertical projections from a 'soft' grid in plan, illustrate the use of axonometrics to describe urban topography and building form at an urban scale. In the upper image note the simple technique of leaving the existing fabric in plan, projecting only the proposed development.

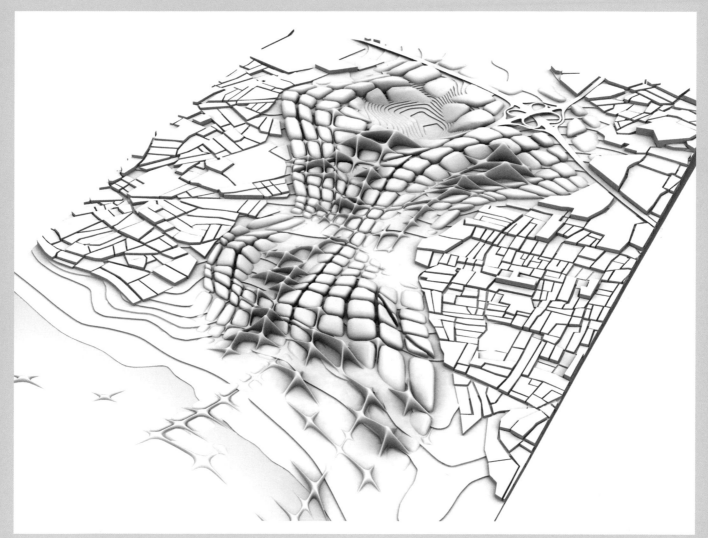

4

5

5. There are parallels between the visual character of the postcard-inspired image and this urban axonometric by Will Alsop. Alsop's axonometric is unequivocal in the way it draws attention to the focus building: the detail of the context is sufficient to describe the textures and scale of the city, but in a way that does not distract from the visual allure of the tower, with its distinctively inventive façade treatment.

6. Stephenson Bell's exploded axonometric shows their proposal for a building on Fountain Street in Manchester. The drawing is an exploded isometric drawing and was produced at presentation stage to convey the holistic design approach taken with the architectural design. The drawing is a pen-on-tracing paper sketch over a static shot of a virtual model that was produced in 3ds Max. The sketch was then scanned and additional information added in Photoshop to produce an image capable of clearly explaining all aspects of the architectural, structural and environmental principals at once. The image was then used in conjunction with renderings taken from the virtual model to form a cohesive presentation.

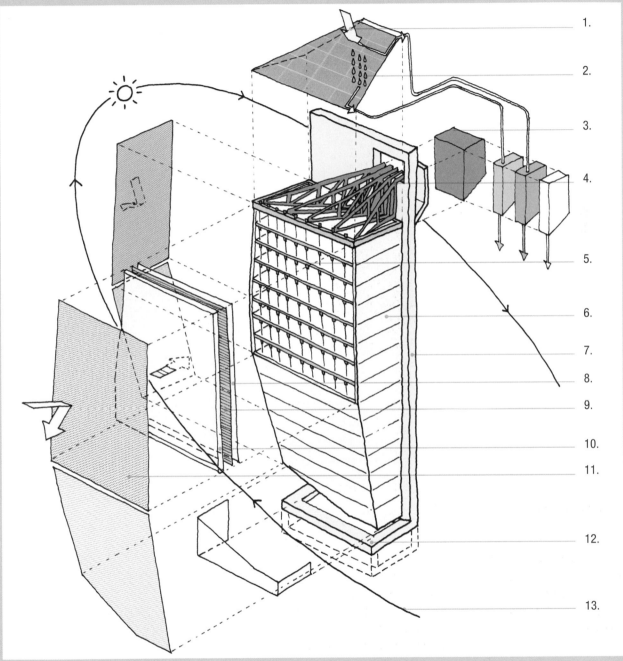

1.

2.

3.

4.

5.

6.

7.

8.

9.

10.

11.

12.

13.

6

1. Solar energy heats landlord hot water

2. Grey water collected for toilet flushing, etc

3. Counterbalance

4. Steel truss supports uppermost floor plate

5. Lower floors suspended by perimeter cables
tied back to uppermost floor (max 14 x 27m
column free floor plates)

6. No fritting to north-facing elevation

7. Limestone-clad core

8. Single-glazed inner 'skin'

9. Blinds within the cavity block low winter sun

10. Double-glazed outer 'skin'

11. Frit pattern on glass prevents excessive solar
gain (frit pattern density varies by facet)

12. Column-free ground floor with double-height
reception space

13. Sun path

STEP BY STEP USING SIMPLE CONVENTIONS

These examples by Chris Staniowski illustrate some of the simple conventions in drawing axonometric and isometric projections. Illustrated below and opposite are steps showing an isometric projection. The opposite page also shows examples of types of axonometric illustration.

TIP CUTAWAYS

Use a cutaway axonometric to explore sequence and time in a project, and to explore otherwise hidden objects and features of your design.

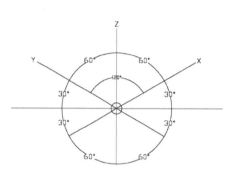

1. Establish the direction of the three principal axes X, Y and Z.

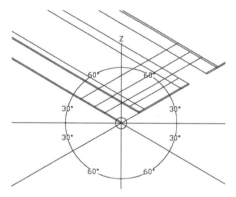

2. Using the plan, measure key points on the drawing and transfer these dimensions along the X and Y axes.

Every line that is drawn should run in parallel to one of the axes.

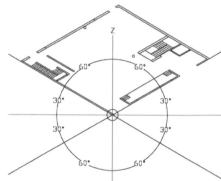

3. Using the basic principle of measuring the plan and transferring the measurements to the X and Y axes, the plan should start to emerge only at the angle defined by the three principal axes.

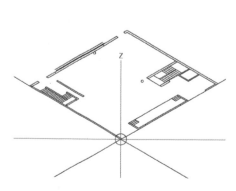

4. Once complete, and using another layer or sheet of tracing paper, begin assigning measurements along the Z axis, drawing lines vertically from each key point on the plan where walls and objects intersect.

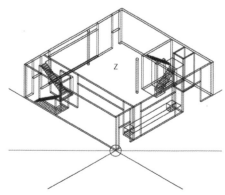

5. You will now be presented with a three-dimensional representation of the building known as a wire-frame drawing. Once this stage is reached the drawing can either be left as a wire-frame model or certain lines can be removed to make each wall solid, obstructing the view of architectural elements within the drawing.

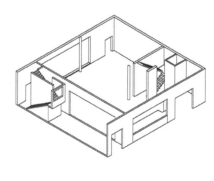

6. Final drawing with a key wall removed.

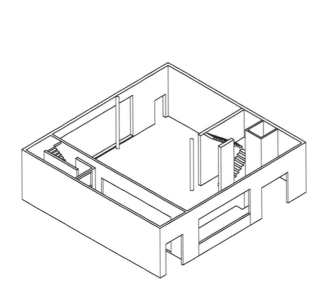

Isometric drawing

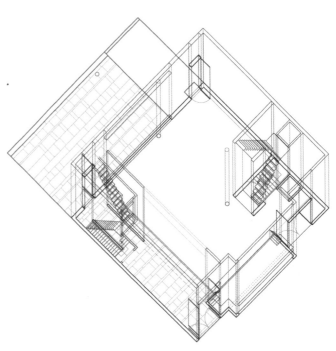

Wire frame axonometric

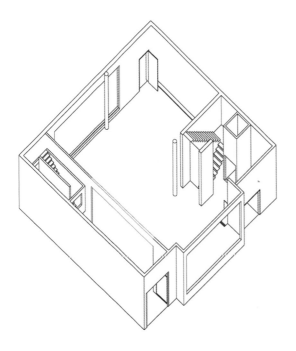

Solid axonometric

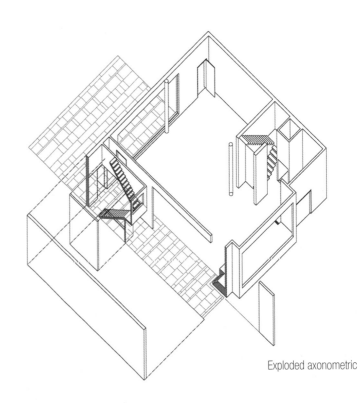

Exploded axonometric

TIP EYE HEIGHTS

Think carefully about eye heights and
how they are translated into horizon
levels; this will affect what is visible in
the image.

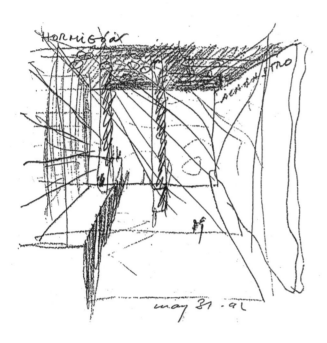

Perspectives

The technique of perspective drawing is an ancient
representational tool used to depict distance or spatial
depth on a two-dimensional plane. There is a rich history
of early perspectival drawing that eventually became
subject to the consistent geometric, and subsequent
mathematical rules that we might recognize today.
Perspective not only opened up new representational
techniques, it was a powerful way to see and to construct,
fostering a new way of thinking and ordering space that
was to become a powerful theme in later Western culture.

Today the persuasive power of perspectives persists:
computer-generated perspective has become the single
most powerful tool to communicate a project. Usurping
the traditional artist's impression, the digital perspective
shows, with unerring certainty, how the project relates
to its context externally and what the spaces will be like
internally. Often photorealistic, these modern perspectives
now play a central role in architectural visualization.

There are two fundamental observations embodied
in perspective. First, that objects appear smaller in the
distance than they do close up. Second, that objects appear
to become foreshortened along the line of sight.

The geometry of perspective drawing starts with the
idea of a picture plane, an abstract plane that is held up
to the object in view; and vanishing points, the points on
the horizon where parallel lines appear to converge. The

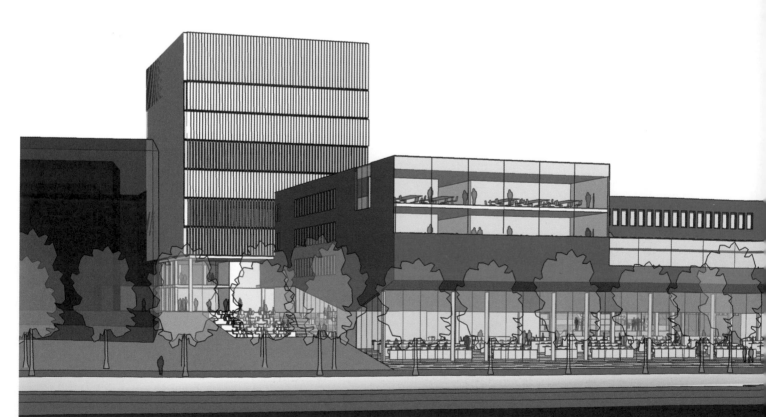

number of vanishing points, which will depend on the viewing angle, will determine what kind of perspective is drawn. These are usually one-, two- or, more exceptionally, three-point linear perspectives.

A one-point perspective is perspective in its simplest form. It can be used when the main lines within the room or building are parallel or perpendicular to the line of sight, for example an interior view of a simple rectilinear room or a head-on view of an external façade.

A two-point perspective, on the other hand, is less static as an image and it allows a greater variety of viewpoints. It allows the viewer to be looking towards a corner of a room say, rather than remaining perpendicular to the main elements of the space. The room will now have two vanishing points (or potentially more) on the horizon; one for each set of parallel lines. When looking towards the corner of a room, the two walls would recede to respective vanishing points.

Finally a three-point perspective, which is more difficult to generate, is used to describe an additional vertical recession, as the building is seen from above or below.

Each of these perspectival grids can be constructed using basic computer modelling software. This facilitates what can be a painstaking process by hand, though at the same time it is also relatively quick to accurately sketch in perspective. Whatever method is used it is important

that perspectives are not presented as the only illustrations of a final proposal, but are shown together with other drawings, as drivers in the creative process of thinking about the building as a whole.

Independently, the final perspective illustration can act more as a marketing tool than a creative contribution to our understanding of the building. New ways of embedding the idea of three-dimensional representations of settings will combine diverse projections with models and film or animation to make up for the over-reliance on one singular geometric construct.

Perspective sketches test ideas. They swiftly reveal a sense of scale, structure and material. Sketching something three-dimensionally, regardless of how accurate the perspective is, opens up new ideas. Even at an early stage, the freely drawn perspective sketch drives the design process as it brings other drawings together; it is a vehicle for synthetic thinking of the project or space as a whole.

TIP USING PERSPECTIVES

Always use perspectives in conjunction with other drawings during the design process. Use perspectives as much to think through as to illustrate a final proposal.

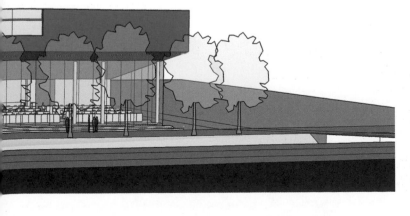

Opposite, top
Alberto Campo Baeza's perspective sketch for the Caja General de Ahorros, Granada, Spain, shows how perspectives can be used as a way to think through early design stages.

Left
This view by Crepain Binst shows the context for their school in Artevelde, Ghent, Belgium. There is an appropriate restraint to the drawing, which is built up from a simple model and transparent layers of tone and shadow.

Case Studies: Perspectives

1

1. A more developed, and particularly effective, hand-drawn perspective of Urbis Prow, Manchester, by Patrick Thomas of Ian Simpson Architects, had to convey building geometry accurately for purposes of a planning application. It was initially drawn by hand in ink on 112gsm A1 tracing paper using a Mars Magno 0.18mm nib pen. It was traced over a crude 'underlay' which involved splicing an early Form-Z model render showing the building's outer skin geometry only (ie no internal or cladding detail) together with a print of a digital camera shot of existing conditions, taken from the same viewpoint. The final traced image was scanned in black and white, and coloured simply using Photoshop.

2

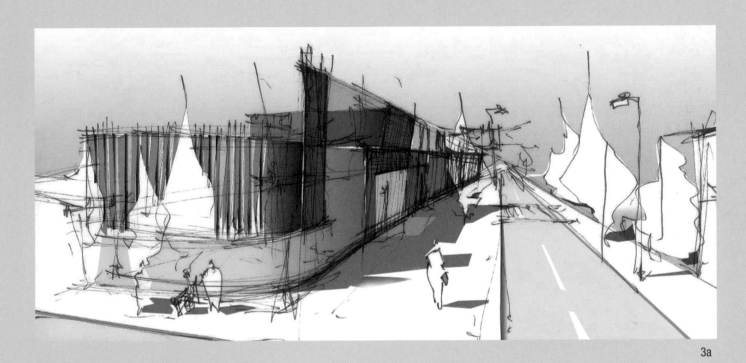

3a

2. Eric Parry's pencil study of the façade of a building in Finsbury Square, London is a wonderful piece of drafting, describing one of London's most contemporary additions in a medium that seems sympathetic to the quality of its masonry. The drawing, done on soft tissue, reads at one level as a study of light and stone. The perspective lines are subdued, creating only a loose framework for the sciagraphy. This reveals the depth of the façade, primary and second orders of piers, and the stanchions that moderate light conditions in the interior and which animate the body of the building.

3a, b. These skilfully executed perspectives by Kyle Henderson demonstrate a presentation drawing based on a one-point perspective, first done by hand and then worked up in Photoshop.

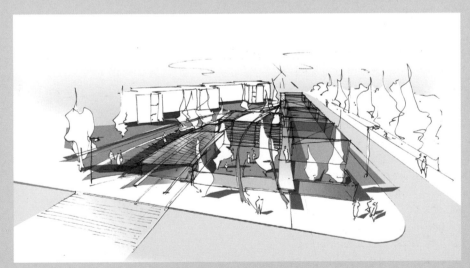

3b

4

4. There is a similar delicacy to Meadowcroft
Griffin's computer-generated perspective. Worked
first as a wire-frame perspective in Microstation,
the image is carefully built up as a series of
delicate transparent layers to emphasize the
interchange between the interior of the building,
external gardens and wider context. This spatial
ambiguity is balanced by precise shadows and
key reflections that establish the body of the
building and orientations.

TIP RENDERING

Only render what is important. Use
render to draw attention to significant
elements in the perspective drawing.

5a

5b

5a, b. Neil Denari's perspective renderings of the Corrugated Duct House (top) and Vertical Smooth House (bottom) are classic demonstrations of digital renderings of architectural design where perspective is set against well-balanced natural and artificial lighting.

STEP BY STEP DRAWING A ONE-POINT PERSPECTIVE

These examples by Chris Staniowski illustrate the way
to create a one-point perspective starting with a plan.

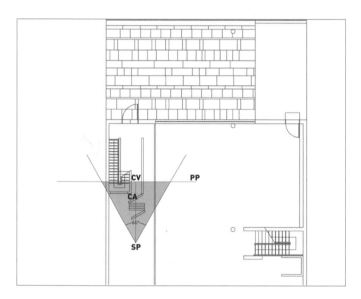

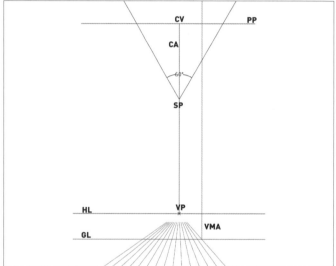

1 Start by looking at the plan. Imagine yourself in the building and think about what you would like to see. Think about the most interesting features that would describe your design to others. What would you see and what would be hidden from view?

The field of view for this drawing is highlighted in grey; note what you will see and where walls will restrict your view. These will need to be made evident in the perspective drawing, as they are a product of the design.

Establish the station point (SP). Simply put, this is where you would be standing observing the perspective you are about to draw. Everything you want to describe should fit within a 60-degree field of vision; if not move the SP backwards or further away from what you want to draw.

Draw a vertical line for the central axis (CA) from the station point and establish a centre of vision (CV) along that line followed by a picture plane intersecting this point. This locates the distant points and objects we will draw in the perspective. Extend two lines at 30 degrees from the CA between the SP and the picture plane (PP). Using the plan set-up as a guide, re-draw the PP as the horizon line (HL), while the CV will become the vanishing point (VP) for the perspective drawing.

2 The ground line (GL) and the vertical measuring axis (VMA) should be drawn an equal distance from both the HL and the CA parallel with each and perpendicular to each other. Along the GL draw equal points of measurement – you may wish to use small increments of measurement on a detailed drawing or larger increments on a less detailed drawing. Since this drawing is relatively simple, the increments are relatively large. Put the same equal measurements on the VMA. These measurements can be extended towards the vanishing point and taken below the ground line. The will give the effect of receding into the distance – the goal of drawing an accurate perspective.

Measurements should be also be added to the CA before proceeding onto the next step.

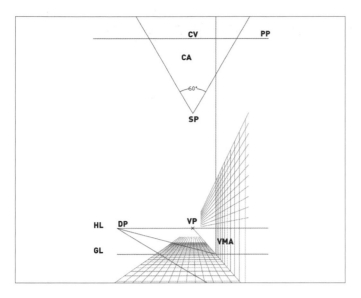

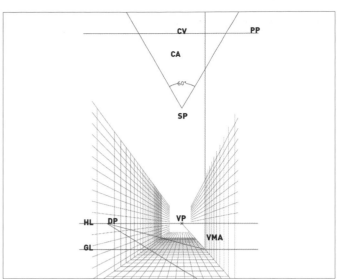

3 In order to complete the perspective grid you need to establish a series of horizontal lines. However, like the vertical lines they too will be subject to the vanishing point in a single-point perspective drawing.

Drawing two 45-degree lines from the SP will intersect the HL and give you a diagonal point (DP). Where this line crosses the grid lines are points at which horizontal lines can de drawn, parallel to the GL. This allows you to complete the grid below the GL. To complete the grid above the GL, draw a line between the intersection of the GL and the VMA and a point halfway between the DP and the VP.

4 At the edge of the grid you have just created you can draw vertical lines and, using the same principles, create a series of grids that define the walls and ceiling. Measurements can be taken from the plans and sections and applied to the perspective grid to ensure accuracy.

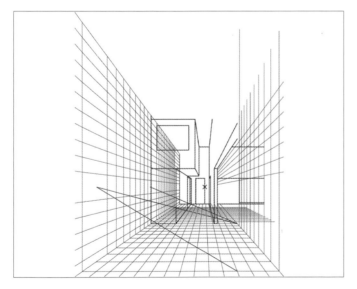

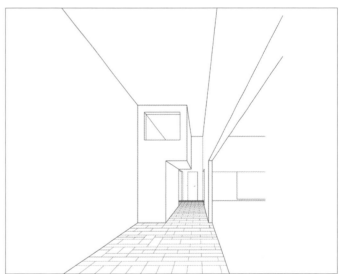

5 This grid defines points in space and allows you to plot points in three-dimensional space quickly and accurately, allowing you to create precise representations of your spaces at a human scale. The perspective grid can be used again for creating other internal and external perspectives of a similar size and scale for your building.

Much like the isometric drawing (see page 122), you can take measurements from the plan and transfer them on to the perspective grid to create the footprint of a chosen object.

The height of objects is determined by measuring off the CA. All non-vertical lines should be seen to converge upon the vanishing points to give the illusion of perspective.

6 People and familiar objects appropriate to the building can be added to the final perspective view in order to give the drawing a sense of scale.

STEP BY STEP DRAWING A TWO-POINT PERSPECTIVE

Here, a two-point perspective is drawn using measuring points on a two-point perspective grid.

Start by looking at the plan. Imagine yourself in the building and think about what you would like to see. Think about the most interesting features that would describe your design to others; what you would see and what would be hidden from view. The field of view for this drawing is highlighted in grey; you will see the staircase, a column, the wall to the left and the wall to the right.

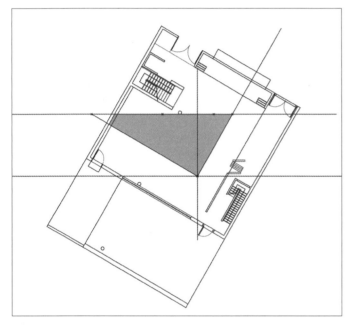

1 Having chosen an interesting view or perspective, draw a plan or a simple representation of the building at an appropriate scale. This should be oriented so that the vertical lines within the field of view can simply be projected downwards from the grid you are about to create. (The reason for this will become more evident later.)

The first and most important point to establish is the station point (SP). Having imagined what you want to see when standing in your building, this point represents where you will be. Most of what you want to describe in the drawing will fall within a 60-degree field of view.

Draw a vertical line from the SP up the page; this is the central axis (CA), followed by the line of the picture plane (PP) perpendicular to this axis. It is useful for the PP to intersect a vertical in the space so that vertical measurements can be easily established and the perspective can be drawn with an accurate representation of height.

Next it's time to establish the vanishing points (VPs). Draw lines between the SP and the PP. It is important that these lines are parallel with the plan. Where these lines intersect the PP establish the left and right VP.

2 Top image. These few points are the key to establishing a perspective grid, which you will use to draw your perspective. The diagram can now be moved vertically down the page so that you can it see it more clearly.

Placing a compass on each VP, draw an arc between the SP and the PP, creating measuring points (MPs) – one to the left and one to the right of the CA. These points are important, as they will establish the grid you will need to construct the perspective.

Now it is time to construct the perspective grid.

3 Bottom image. Establish the horizon line (HL), which represents your eye level in the drawing followed by the ground line (GL). The GL is nothing more than a line upon which measurements are laid out but it must intersect the SP.

Transfer the MPs and the VPs to the new HL.

Along the GL draw equal points of measurement; you may wish to use small increments of measurement on a detailed drawing or larger increments on a less detailed drawing. This drawing is relatively simple so the increments are quite large. Put the same equal measurements on the CA.

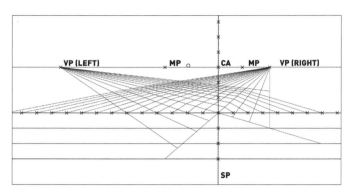

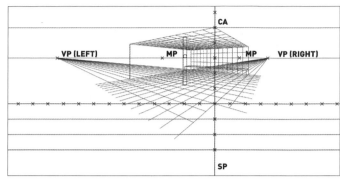

4 Between each point of measurement on the GL and the measuring point on the opposite side of the vertical axis draw lines to the line between the VPs and the SP. Repeat this process on the opposite side.

You should see a new set of measuring points on the lines that establish the field of vision. Draw lines between these new points and the VPs to create the perspective grid.

If necessary these points can be further subdivided to create a more detailed grid.

5 The process can be repeated to produce an overhead grid at the required ceiling height and to the sides to create a measuring grid along a wall.

This grid defines points in space and allows you to plot points in three-dimensional space quickly and accurately, allowing you to create accurate representations of your spaces at a human scale.

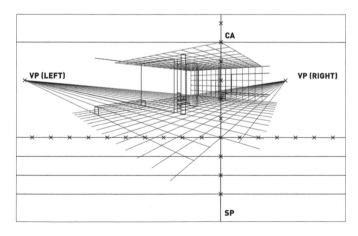

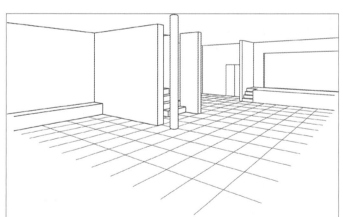

6 The perspective grid can be used again for creating other internal and external perspectives of a similar size and scale for your building.

Much like the isometric drawing (see page 122) you can take measurements from the plan and transfer them on to the perspective grid to create the footprint of a chosen object.

The height of objects is determined by measuring off the CA.

All non-vertical lines should be seen to converge upon the vanishing points to give the illusion of perspective.

7 Add people/familiar objects appropriate to the building to give a sense of scale.

STEP BY STEP MAKING RAPID PERSPECTIVE SKETCHES BY HAND

The illustrations here show a method of rapidly making one-point perspective sketches.

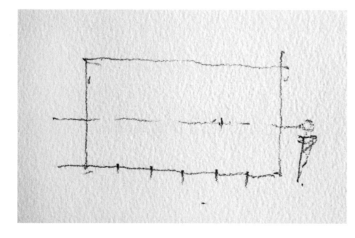

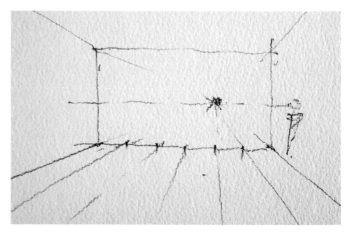

1 Start with an elevation – either internal or external. This should be sketched in proportion so that running dimensions can be estimated (usually) along the lower edge.

2 Estimate the eye level. Standard heights are 1.5m for standing and 1.35m for sitting. Scale off the elevation to draw a horizon line (HL) at the chosen eye height.

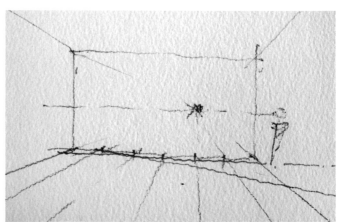

3 Locate the vanishing point (VP) on the HL. If necessary a set of vertical scales can be marked off along edges of the elevation. Project from the VP to establish perspectival lines.

4 The next stage of the drawing is to establish perspectival recession. This is done using the idea that a 45-degree diagonal running across the drawing will locate all the 'squares' in the drawing. This diagonal can be quickly located by eye, as shown: locate a horizontal line just below the lower edge of the elevation. This is an estimate of the extent of the square closest to the elevation.

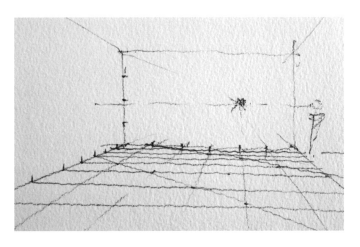

5 A diagonal can then be drawn across the squares to establish all others in plan.

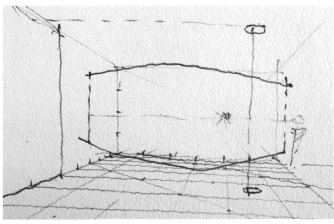

6 Don't overwork the drawing by hand. This kind of sketch can only establish the overall proportions.

7 A perspective sketch can be quickly taken back into Photoshop to add layers of tone or texture, figures and key lines to aid definition.

8 The final artwork.

STEP BY STEP CORRECT PERSPECTIVE DISTORTION

The sequence illustrated here (by Ian Henderson) was created
in Photoshop, but this work could also be done in several
other similar software packages and applications.

2 Set up guidelines. To correct the perspective in the photograph, first set up
some key guidelines or turn on the user grid. To create a guideline, ensure
that the rules are turned on. Using the Move tool, drag a guideline from a
rule and position it.

1 Original photograph. Owing to the wide-angle lens used when taking
this photograph the buildings appear to be leaning back and converging
towards their roofs.

3 Transform image using Distort. To enable you to adjust the image it needs
to be unlocked from the background. Double left-click on the 'background'
layer to convert it into a standard usable layer, and rename it. Ctrl-T to activate
the Transform tool. Right click over the photograph to access the Transform
functions. Select Distort and adjust the anchor points until the perspective has
been corrected. Double click or hit Return to apply changes.

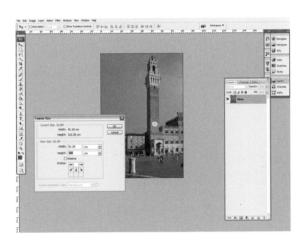

4 Adjust canvas size. As a result of the transformation the people in the lower half of the photograph appear to be squashed. Activate Adjust Canvas Size and increase the canvas size to the bottom of the photograph allowing enough space to correct the proportion of the people.

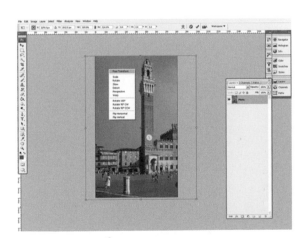

5 Transform image using Free Transform. Ctrl-T to activate the Free Transform tool. Select the lower mid-anchor point and stretch downwards until the people in the photograph appear in correct proportion. Crop any leftover canvas using the Crop tool.

6 Final image. The buildings in the final photograph now have perfectly vertical lines.

STEP BY STEP CREATING A LATHE MODEL

The following model sequences (by Ian Henderson) can be
created using most generic computer graphics programmes,
such as 3ds Max, Maya, Softimage or LightWave.

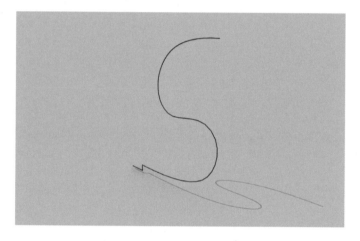

1 Spline. Create the required shape using the Line tool.

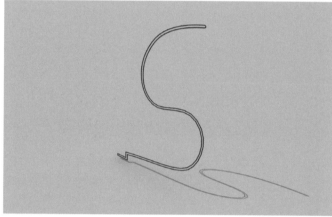

2 Spline shape section profile. Adjust the spline to create a section profile. In
this case the spline was offset using the Edit Spline Outline tool to provide
the section profile with some thickness, and two vertices were filleted to
round off the end of the profile.

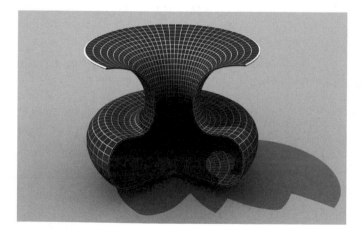

3 Creating a lathe mesh. Select the section profile spline. Create a lathe mesh
using the 'lathe' modifier from the modifier list.

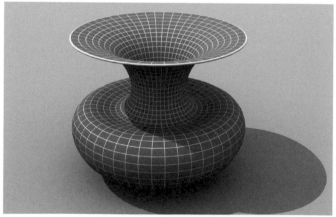

4 Lathe mesh. Adjust the 'lathe' modifiers parameter settings to refine the mesh.

5 Mapping a material. A material was assigned to the surface. The mapping
coordinates were set using the lathe meshes' in-built mapping system.

STEP BY STEP CREATING A LOFT MODEL

1 Spline path. Create a path from a spline. In this case the 'helix' was used.

2 Spline shapes. Create spline shapes. In this case a 'circle' and 'star' were used.

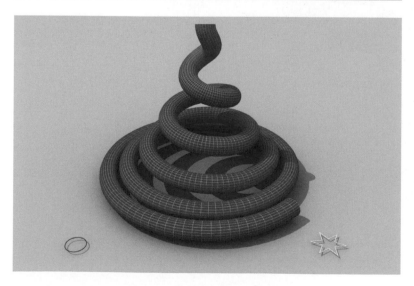

3 Loft circle. To create a mesh object first select the path spline. From loft parameters under 'creation method' add the 'circle' spline shape using the Get Shape button. The order in which the splines are selected is crucial. If the path spline is selected first and the shape spline added subsequently, the mesh created would be located exactly where the path spline is. If the shape spline is selected first and the path spline chosen subsequently, the mesh created will be located where the shape spline is.

4 Add star to loft. To form a more complex mesh any number of other shape splines can be added to the loft. To add another shape spline first determine where you want the shape spline to appear along the path using Path Parameters and simply add the additional shape spline using the Get Shape button.

In this case Path Percentage was used and the star shape spline was added at 100% (the end of the path spline).

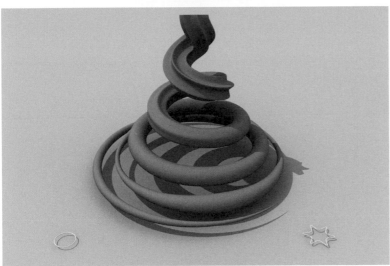

5 Refine mesh. To refine the mesh further the positions of the shape splines were adjusted and the shape splines scale was adjusted at various points along the path spline using the 'scale' deformation graph.

To refine and position the blend between the circle shape spline and the star shape spline, the circle shape spline was placed at 0% and 80% and the star shape spline at 100% along the path spline. This resulted in the blend between the circle shape spline and the star shape spline occurring in the last 20% of the path's length instead of the blend occurring along the full 100% of its length.

The star shape spline was modified to produce a softer-looking mesh by filleting the stars' points and adjusting the 'fillet radius' under the star shape spline's parameters.

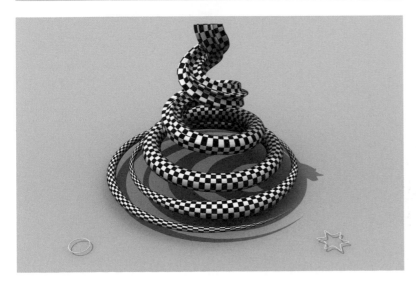

6 Mapping a material. A material was assigned to the surface. The mapping coordinates were set using the loft meshes' in-built mapping system.

STEP BY STEP POLYGON MODEL EDITING

A Polygon mesh can be edited on a sub-object level (referred to as 'sub-object editing'). The key principle behind this form of mesh manipulation is that the modeller has access to the basic structural components that form a mesh. These components consist of vertices, edges and polygons that can be manipulated to alter the form the mesh.

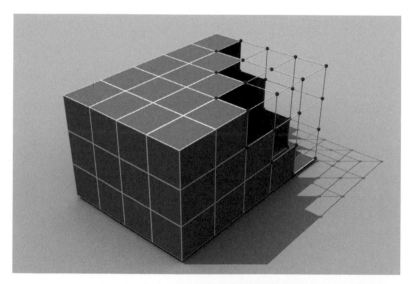

1 Vertices. A vertex is the most basic element of a mesh. Vertices are single structural points that define the corners of each face within the mesh.

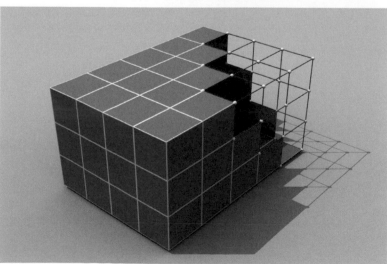

2 Edges. Edges are the second most basic element of a mesh. Edges tie the vertices together to form the edges of faces within the mesh.

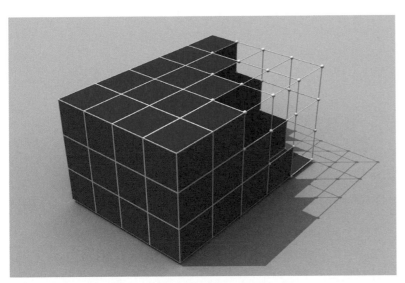

3 Polygons. Polygons are formed by connecting vertices and edges together. Polygons are a series of faces that are connected to form a mesh.

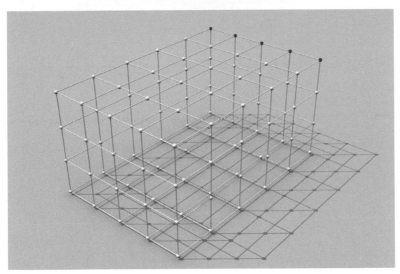

4 Select vertices. Adjusting its vertices can modify a mesh. The desired vertices must first be selected.

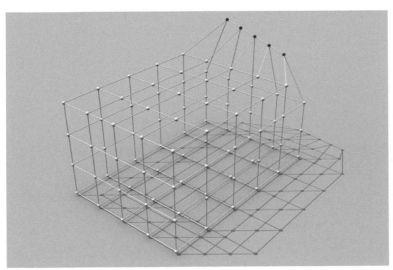

5 Edit vertices. The selected vertices can be manipulated using the Move, Rotate and Scale tools. Additional editing tools are available within the Modify tab for manipulating the mesh's vertices.

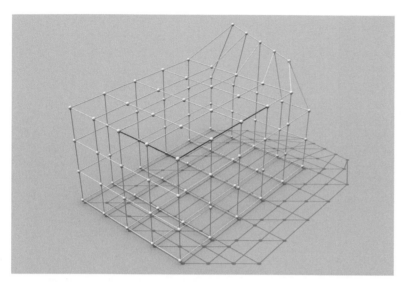

6 Select edges. Adjusting its edges can modify a mesh. The desired edges must first be selected.

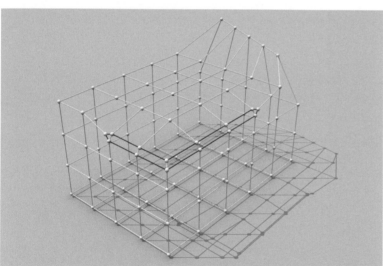

7 Edit edges. The selected edges can be manipulated using the Move, Rotate and Scale tools. Additional editing tools are available within the Modify tab for manipulating the mesh's edges. In this case a Chamfer tool was used to chamfer the selected edges.

8 Select faces. Adjusting its polygons can modify a mesh. The desired polygons must first be selected.

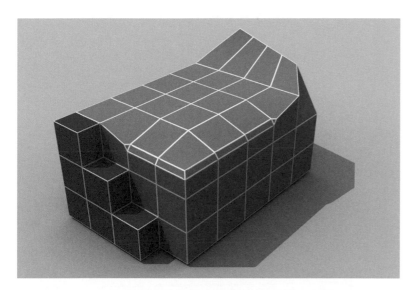

9 Edit faces. The selected polygons can be manipulated using the Move, Rotate and Scale tools. Additional editing tools are available within the Modify tab for manipulating the mesh's polygons. In this case the polygons were extruded using the Extrude tool.

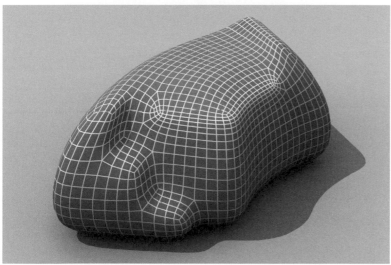

10 Using modifiers. Modifiers can be applied to the mesh to further alter its form. In this case the 'turbo smooth', 'bend' and 'taper' modifiers were used to further manipulate the mesh's form.

11 Mapping coordinates. A material was assigned to the surface. The 'UVW map' modifier was used to map the material correctly on to the surface.

STEP BY STEP SOLID MODEL EDITING

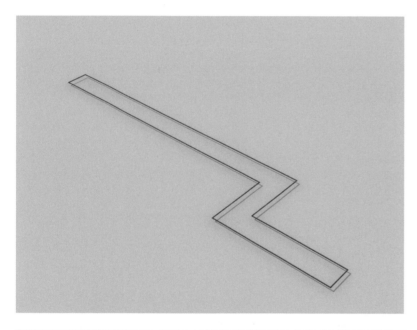

1 Spline shape. Create a spline shape using the Line tool.

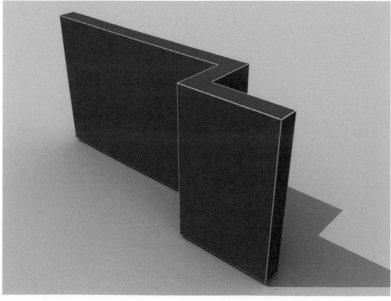

2 Extrude the spline. Extrude the spline to make a solid mesh using the 'extrude' modifier from the modifier list.

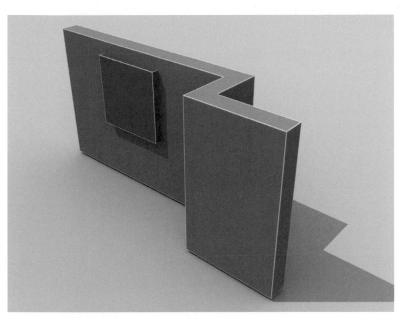

3 Position mesh object to subtract. Create a smaller mesh object to subtract from the large mesh. In this case a box was created from the standard Primitives menu. Position the smaller mesh object using the Move tool.

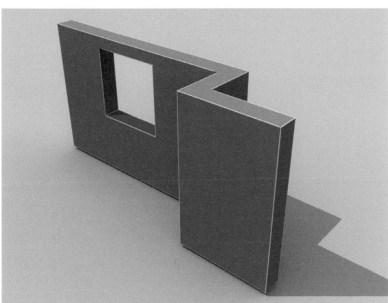

4 Boolean operation. Select the large mesh to be retained. Use the Pro Boolean tool, then select 'subtraction' from the Boolean parameters and pick the small mesh object. Ensure there is a good overlap between both mesh objects to ensure a clean cut is made.

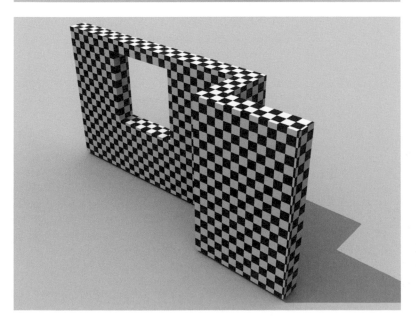

5 Mapping a material. A material was assigned to the mesh object. The 'UVW map' modifier was used to map the material correctly on to the mesh object.

STEP BY STEP CREATING A SPLINE SURFACE MODEL

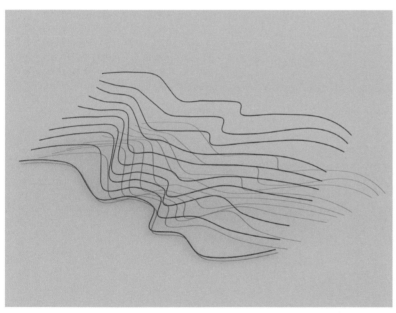

1 Splines. Create required spline shapes using the Line tool. For multiple splines attempt, where possible, to use an equal number of vertices on each spline. Create one spline object by attaching all the splines together. When attaching the splines select the lowest spline and attach each spline in order of sequence.

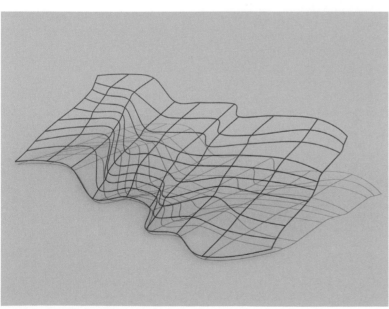

2 Spline cage. Create a spline cage by connecting the vertices with new splines to form a closed lattice made up of three- or four-sided shapes. The 'cross-section' modifier can be used, or connect the vertices manually using the Create Line tool from within the 'edit spline' parameters.

When creating the splines manually ensure the vertices align properly by using the 3D Snap function.

When using the 'cross-section' modifier the first vertices of each spline should all start at the same side to ensure the modifier works correctly.

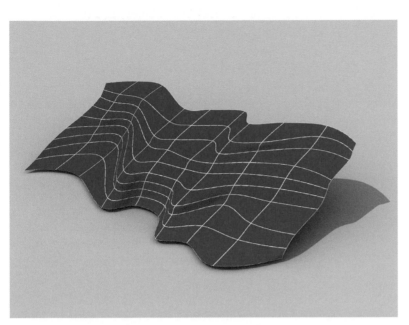

3 Surface. Use the 'surface' modifier to create surface mesh over the spline cage.

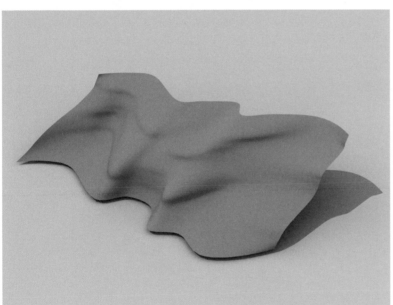

4 Surface final. Adjust the parameters for the surface modifier to suit. The spline cage must be made of a three- and/or four-sided grid to prevent any holes appearing in the surface mesh. At the junction where two splines cross, the two respective vertices must align exactly to ensure the surface mesh is created properly.

5 Mapping a material. A material was assigned to the surface. The 'UVW map' modifier was used to map the material correctly to the surface.

Scripted drawings
Matthew David Ault

Design techniques that employ scripting look to mathematical rules to generate formal solutions to complex design problems. To do this, the design process can be broken down into an algorithm. The algorithm can be understood as a procedure by which something is made, a sequence of instructions or a method statement. Algorithmic design allows us to model concepts. Modelling in this way encapsulates information about the nature of a design in which relationships are relative.

The following case studies illustrate examples of algorithmic approaches to architectural design and the role of drawing within them. When using algorithms the designer is designing the methods and processes.

First the algorithm must be converted into a language or code, called a script, which can be understood by the computer. There is often a script editor within 3D visualization software that records or displays the sequence of commands used to create forms on screen. By using this script and adding variables and loops, more complex objects or arrangements can be created.

The script can be modified to place any object or component in an array, for example a cylinder or sphere. Following the same principle, instead of proliferating the component across a Cartesian grid, the script could instead be applied to a non-Euclidean surface, acting as a

scaffold across which the component is distributed.

Instead of placing the component at particular points in space, the script could specify a location on the surface, the UV coordinates or parameters. The advantage of this is that the specific shape of the surface and its position in space is not needed.

In example 1 (below left) a cube is created, scaled and moved. This sequence is repeated several times, each time moving a new cube further. In 2 (below centre) the cube is also rotated each time it is created. In 3 (below right), a second loop is added so that an array of cubes is created.

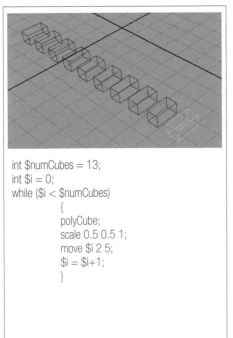

```
int $numCubes = 13;
int $i = 0;
while ($i < $numCubes)
        {
        polyCube;
        scale 0.5 0.5 1;
        move $i 2 5;
        $i = $i+1;
        }
```

1

```
int $numCubes = 13;
int $i = 0;
while ($i < $numCubes)
        {
        polyCube;
        scale 0.5 0.5 1;
        move $i 2 5;
        int $rotation = $i*15;
        rotate 0 $rotation 0;
        $i = $i+1;
        }
```

2

```
int $numCubes = 13; int $numRows = 10;
int $i = 0; int $j = 0;
while ($i < $numCubes)
{
        while ($j < $numRows)
        {
        polyCube; scale 0.5 0.5 1;
        int $rotation = $i*15+$j*10;
        move $i $j 5; rotate 0 $rotation 0;
        $j = $j+1;
        }
$j=0; $i = $i+1;
}
```

3

Simple scripts can deal with repetition and complexity.

1 Stage One
2 Stage Two
3 Stage Three

In these illustrations, the orientation of a component is scripted to follow the position of the Sun over time. Thus the assembly could be used as a tool to explore the implications of the surface morphology on shadow patterns and by programming at different times of day the method could be used to determine the requirements for, say, a canopy structure.

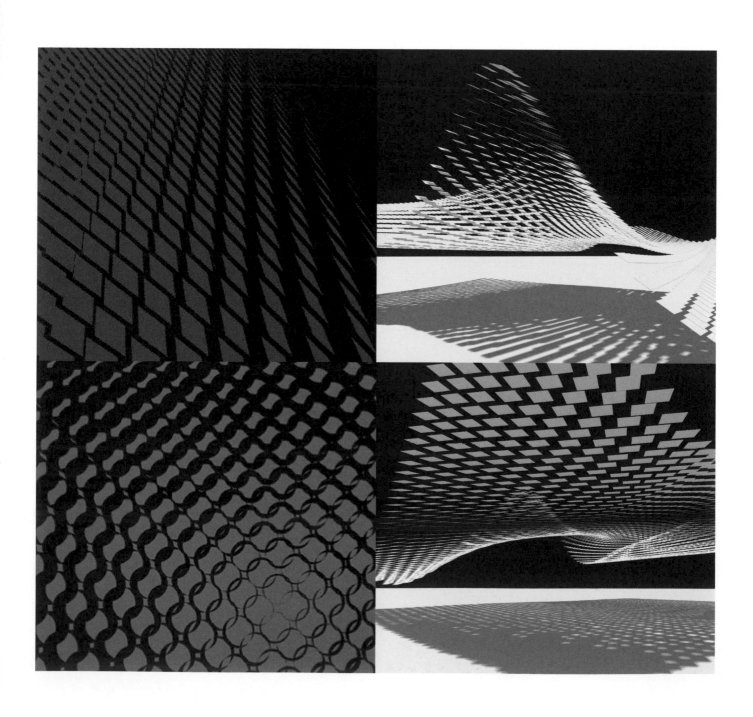

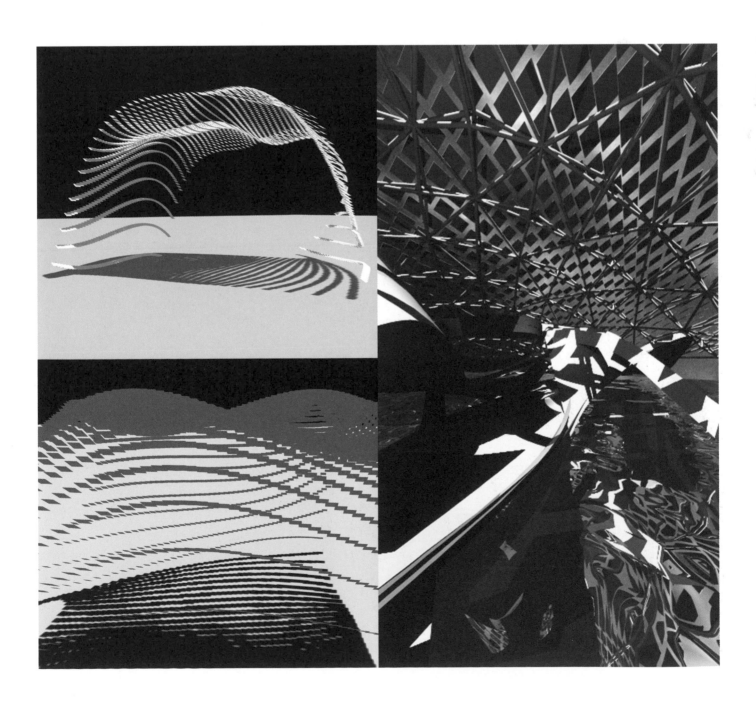

In addition to scripting, relationships can also be established using modelling techniques. Digital modelling is distinguished from computer-aided drafting in that the geometry is not drawn in a serial fashion, but arises from a set of relationships in an interconnected model of a design concept.

An understanding of the underlying nature of the idea is required – not simply its visual appearance – and so the rough sketch idea (below left) is a useful starting point.

The components that were proliferated over a non-Euclidean surface (see pages 153–154) were all identical. However, component morphology can be defined to respond to that of the scaffold surface, the local component geometry being affected by the global surface morphology.

The sketch shows how the geometry of the

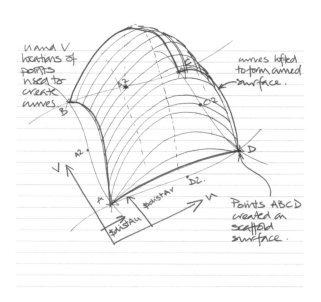

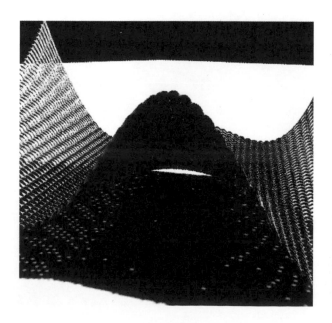

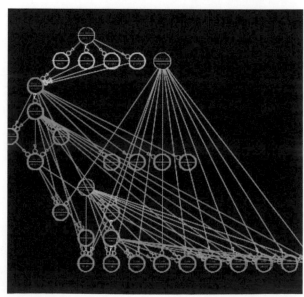

component is not absolute, but relative to the surface. In a hierarchical way, the component cannot be described without reference to the surface it lies on and its location thereon.

The symbolic diagram (bottom centre) shows the relationships between elements of geometry required to define the simple structural like component below.

Four vertices of a polygon define the component. Because the geometry is defined as relationships and not as numeric values, any change in the location of the vertices will affect the geometry.

The component is proliferated across a surface subdivided into polygons (bottom). Each instance of the component within four polygon vertices will vary according to the size of the polygon that defines it.

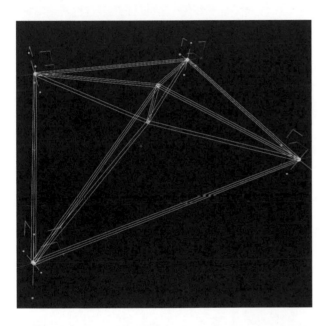

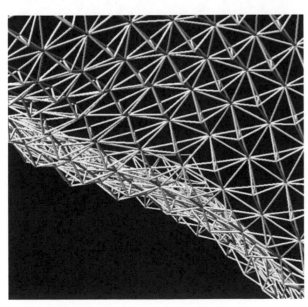

Designing using associative and parametric modelling allows the incorporation of information such as material limits that may constrain possible forms. It also aids the incorporation of analysis into the development and improvement of the initial concept; the designer can adjust parameters and make decisions based on the feedback. This is not always to 'optimize' forms, for example in terms of structural performance, but may relate to other aspects or roles required of the design. Prototyping and manufacturing should form an integral part of these design processes, for example three-dimensional prints or scale prototypes using laser cutters.

The design process begins with ideas that define the performance priorities and requirements and

Below
The images below show the progression of a design from its initial sketch geometry, definition of parameters and variables, digital modelling, manufacturing and anaylsis.

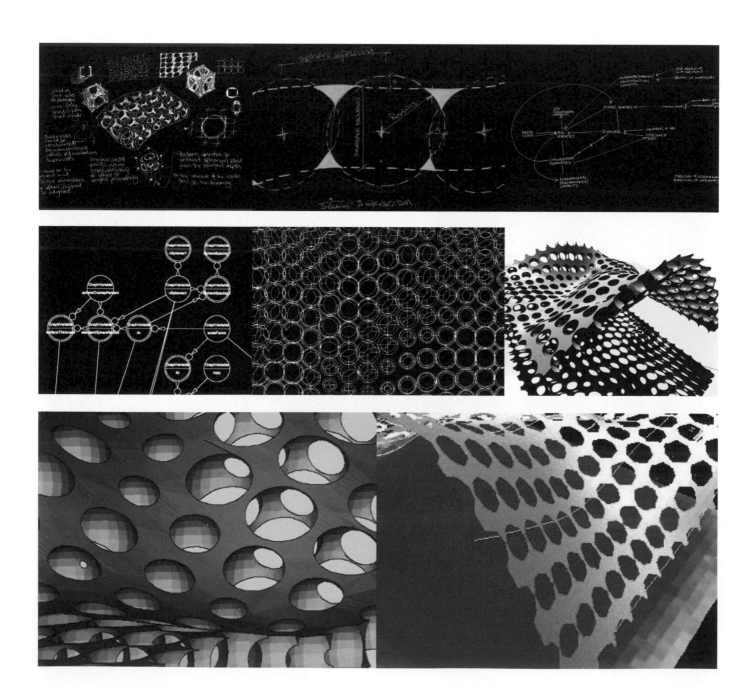

manufacturing constraints. These are then translated into a geometry statement or relational/hierarchical diagram. This ensures that any future testing of the idea through modifying characteristics or parameters, perhaps as part of an evolutionary algorithm, can be accommodated.

Here an academic exercise explores how a porous material might act as a shading device. The shadows cast and the lighting effects beneath the 'canopy' were investigated.

Surface curvature creates modulation in shadow patterns, varying with the movement of the Sun. The integrity of the system breaks down when curvature increases too greatly, and therefore limits had to be included to address this. These related to the size and

intersection of the voids, and the variation in the size of the voids with increasing curvature and thickness of the canopy.

Models were manufactured in plastic using laser sintering and in gypsum powder on a 3D printer. As can be seen in the image, the limits were not correct and the gypsum model did not have sufficient strength to support its own weight (bottom).

Below
These images show physical models produced to inform decisions made at various stages in the design and manufacturing process.

The screen shots below were taken during the testing, evaluation and improvement of the model.

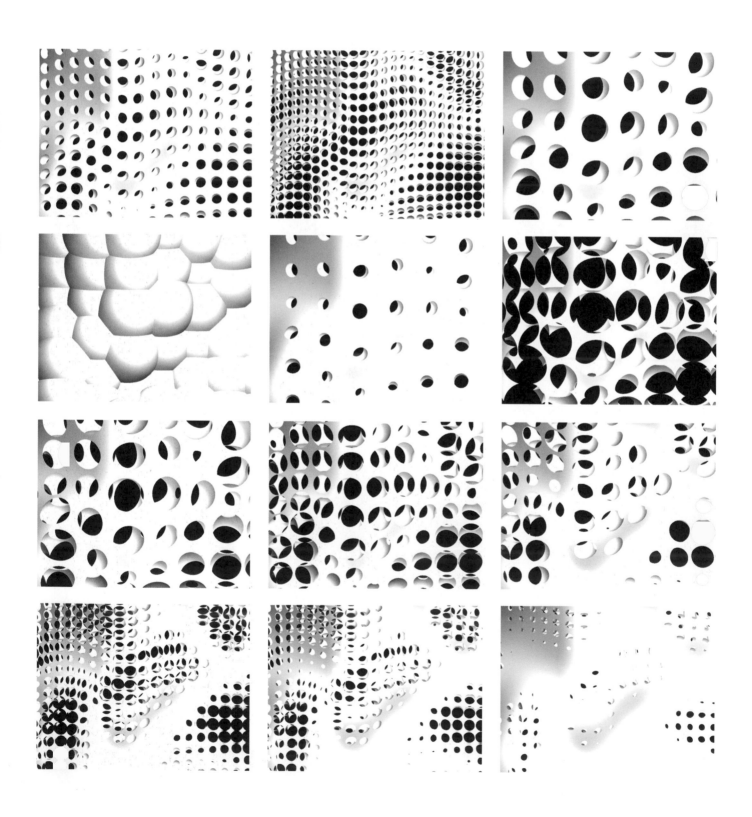

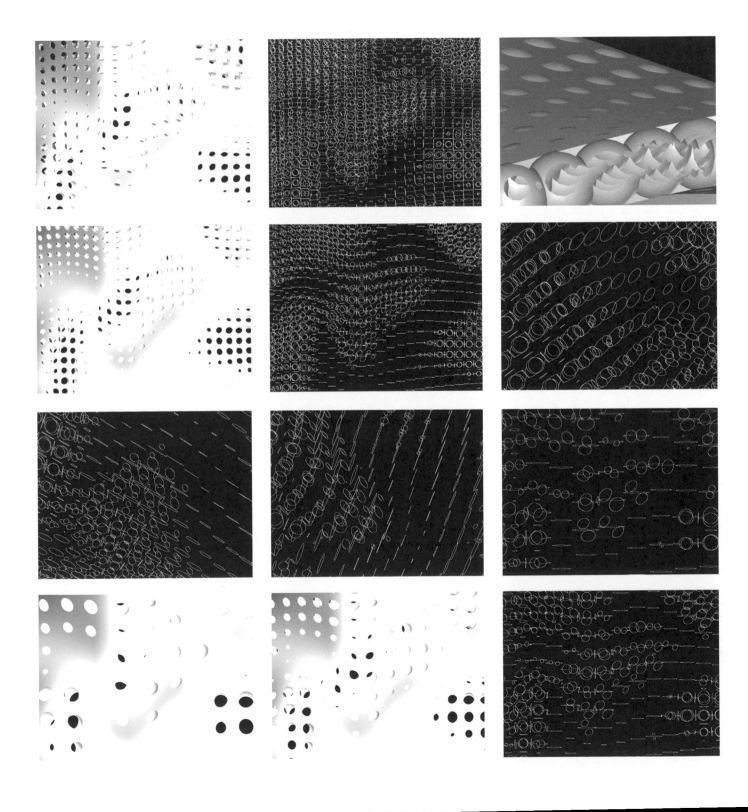

PLACES

Introduction

This section brings together a range of drawing techniques and drawing types, all of which have been explored in earlier parts of the book. It arranges drawings into three clusters, according to the kinds of places they represent: interiors, landscapes and urban settings.

While acknowledging that this simple grouping is reductive, in the sense that it overlooks other kinds of drawings and places, it is nevertheless useful in terms of being able to establish a broad framework for the idea that architectural drawing is ultimately concerned with the creative exploration of a place and the communication of the qualities of that place. This section is about architectural drawings that represent a place where something happens, that has a physical presence, an orientation and a dialogue with its context.

In a sense, bringing together the variety of drawing techniques and drawing types in this way is an important counterpoint to seeing architectural drawing in terms of aesthetics alone. Unlike fine art, an architect's drawings have a necessary and often pragmatic purpose; they are grounded in the physical world, and are artefacts that construct a transition between the formal and the material imagination, and the eventual constructed place for human events.

An obvious consideration in the way we develop ideas for architecture is the physical and cultural context in which we are working; making images that are concerned with place differentiates architectural drawings from those of the engineer – for the architect the specificity of place matters.

Consequently, a great number of architectural drawings are concerned with physical context and the parts of this section that deal with urban drawings and landscape drawings use computer-generated or physical models extensively to show how the proposed buildings relate their surroundings.

It is a more difficult question to visualize how the building responds to a specific cultural context, but this is where architectural representation needs to look again at stage or film set design, which often create a world that reflects non-visible aspects of the narrative.

Interiors

The interior is a significant genre in architectural representation. These drawings are distinct from the drawings of the architect–decorator, which have a different focus; drawings of architectural interiors tend to situate internal spaces as part of a sequence and aim to understand the building as a whole with respect to the urban/landscape context and its culture. Historically, drawings of interiors were not represented in terms of an independent aesthetic. Rather the architectural

Previous page

Sara Shafiei (Saraben Studio).
Magician's Theatre, National Botanical Gardens, Rome. Plan view of tracing paper and hand drawing. The design for a Magician's theatre readdresses the sensuality and ornamental richness of the Italian baroque. Notions of magical illusion and geometric anamorphosis generate surgically constructed laser-cut drawings and models that describe the functional solution of the circulation, as well as the special complexity of this realm of projections, performances and illusions.

TIP SPACE

Concentrate on understanding interior space through light and material surface. This will lead on to textures and colour.

interior was traditionally more like a canvas, a container for paintings, frescoes and hangings that represented a matrix of symbolic references, pointing to the physical and cultural context and its historical depth.

Such interiors are replete with distortions and multiple viewing angles and are almost collage-like in the manner in which they first hold and then open space, mediating between the symbolic interior and external landscape or city. Non-perspectival interiors emerge again in modern times most clearly with collages. As we have seen, the collage can be used to bring together more complex sets of relationships that may inform an interior. In this sense the Canadian photographer Georges Rousse is relevant here. Rousse's final product is the photograph, but the interchange between interior/photograph and drawing is a creative journey that opens up unexpected relationships and challenges the dominance of our perspectival understanding of interior.

Like the etchings of Giovanni Battista Piranesi or the watercolours of Joseph Michael Gandy, Rousse's photographs reveal the deep significance of light in our representation of interiors. Visually, interior depth is represented in a drawing by increasing the darkness of the image as the space recedes. In working a drawing by hand this structure tends to be established first, in mid tones. Without it the more detailed elements of the drawing are more difficult to balance. Lights and darks are worked into and out of this basic structure.

The way light works in more detail in an interior is a complex phenomenon. By hand it is possible to swiftly establish a broad structure, but computer-rendering software can more quickly take the depiction of light towards photorealism, as radiosity algorithms now allow for multiple reflections of light off a room's surfaces. These packages can be useful both at the design development stage and as an illustrative device.

Left
Gillian Lambert's detail rendering of the House at Galleon's Reach is evocative of intermediate, partially enclosed space.

Case Studies: Interiors

1

1. This is a preliminary pencil and crayon study for the vestibule that gives access into the Crypt space of St Martin-in-the-Fields, London by Eric Parry. Here the interior is understood as setting up a relationship with the church, the view to which provides an essential orientation during the journey to the underground space.

2a

2b

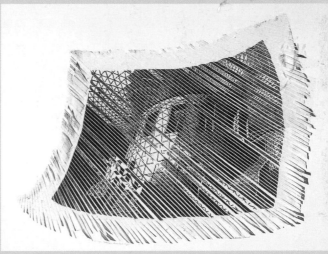

2c

This group of prints is a creative interpretation of the Great Court at the British Museum by the artist Anne Desmet and explores how light works in the space.

2a. *British Museum Series No. 1* is a wood engraving printed in black ink on Japanese Gampi Vellum off-white paper. This image was developed from photographs taken by the artist in daylight hours. The desired image was scaled up on a photocopier to the size of the end-grain boxwood engraving block and the outline structure traced from a mirror-image photocopy on to the block using a selection of fine engraving tools. In the cutting, the artist exaggerated various light effects observed in the building – in this instance, to suggest an idea of how the building might look at night with only one directed light source from one side – and also enhanced the intensity of the shadow fretwork cast onto the curving perimeter wall of the library by the lattice roof.

2b. *British Museum Series No.4* is also a wood engraving printed in black ink on Japanese Gampi Vellum off-white paper. This image was a reworking of the engraved block for *British Museum Series No. 1* and there were a further two developments of this same block before the final stage was reached. The engraving already on the block was reworked and further engraved using a selection of fine engraving tools. In the cutting, the artist exaggerated various light effects observed in the building – in this instance, to suggest an idea of the silvery light of the building in bright daylight. She retained the area of deep shadow along the right-hand side of the curved library building to give a sense of the structure's three-dimensionality within the Great Court space.

2c. *British Museum Diagonal Light* is a wood engraving collaged on card. This collage was made by first printing an impression of *British Museum Series No. 4* on to semi-transparent, buff-coloured, Japanese Kozu-shi paper. Once dry, the print was carefully measured into thin diagonal strips of equal width (each strip being just 3–4mm wide), which were carefully cut out using a sharp scalpel and a steel rule – but leaving a very thin, uncut edge so that the print was sliced but still all in one piece. The image was then laid on to card and glued down allowing a tiny gap to open up between each slice to suggest shafts of light slanting down across the building. To enhance this effect, a few of the print 'slices' were cut out completely, reversed and transposed from left to right to make use of the light effect generated by the paper being semi-transparent, which allows the printed image to show through on the reverse side in a silver-grey tone, in contrast to the stronger black printing on the front. The idea was to experiment with some of the effects of Op Art and also to explore the strong sense of interrupted light within the Great Court.

STEP BY STEP REPRESENTING AN INTERIOR THROUGH SKETCH, LINO CUT AND COLLAGE

1 *Pantheon, Rome (drawing)* by Anne Desmet. Pencil and grey wash sketchbook drawing, made *in situ* in the Pantheon in Rome.

2 *Pantheon.* Two-block lino-cut: one block cut with the key-image that was later printed in blue-black ink; a second block cut to print a pale cream tone over selected highlights of the print, once the first block had been printed. Printed on Somerset Satin white paper. This image is a development from the *Pantheon, Rome (drawing)*, above. The image was scaled up and reversed on a photocopier to the size of the lino blocks and the outline structure traced from a mirror-image photocopy on to the first block. Various perspectival errors in the initial drawing were rectified in the drawing on the block. The image on the block was cut using standard 'V' and 'U' lino-cutting gouges as well as some finer engraving tools and an etching roulette wheel.

3 *Pantheon (tondo).* Lino cut prints and collage on paper. This collage is constructed from 16 roughly triangular strips from spare printings of the *Pantheon* lino-cut, glued together in a circular format, on a rust-red background paper, and with additional yellow colour added by rolling printing ink with a printing roller directly on to the surface of the collage.

The collage was attached with a glue stick and strengthened at the back with lengths of self-adhesive paper tape. The idea of this work was to convey a sense of how it feels to stand in the centre of this extraordinary interior in Rome and to create a sense of being surrounded and enveloped by the curving space.

STEP BY STEP LIGHTING AN INTERIOR USING 3DS MAX AND V-RAY

This sequence by Ian Henderson illustrates the process of changing light effects in an interior image.

1 Main exterior light source. Set up a 'direct' light to act as the Sun and position to provide pleasing rays cast through exterior wall apertures. A 'direct' light is used because rays cast from this light are parallel, thus best replicating the Sun's rays and the formation of shadows cast.

Adjust the 'direct' lights multiplier value to suit the scene.

Use V-Ray shadows for cast shadow type.

Increase the 'subdivs' value under the V-Ray Shadow Parameters tab to improve the quality of the shadows cast and reduce noise.

2 Exterior ambient light. Additional lights are required to simulate ambient light entering the interior. Outside each wall aperture a V-Ray rectangular area light was placed. Physical dimensions of the light should be big enough to cover the extents of the aperture. Increase subdivisions of light to improve quality of shadows and reduce noise. Adjust multiplier value to suit.

3 Bounced light. To simulate bounced light use global illumination. Adjust quality to suit scene and image size. Test using the different Irradiance Map presets, beginning with the lowest quality first.

First bounce was calculated using an Irradiance Map and was set to 'medium'.

Second bounce was calculated using Direct Computation.

4 Environmental light. To simulate additional daylight entering the space via the wall apertures the V-Ray GI Environment and Reflection/Refraction Environment light was used to provide more ambient illumination. Adjust multiplier value to suit scene.

5 Interior lights. Interior artificial lights were added to illuminate specific areas of the scene. Standard spotlights were used with V-Ray shadows.

6 Interior artificial lights were added to illuminate specific areas of the scene. Standard spotlights were used with V-Ray shadows.

Additional lights are required to simulate ambient light entering the interior.

To simulate additional daylight entering the space via the wall apertures the V-Ray 'GI Environment and Reflection/refraction environment' light was used to provide more ambient illumination.

Adjust multiplier value to suit scene.

To simulate bounced light use global illumination.

Adjust quality to suit scene and image size. Test using the different Irradiance map presets to start with beginning with the lowest quality first.

Landscapes

This section covers drawings of architecture that are situated in natural landscapes; architecture intrinsically tied to the land or the gardens it frames. These relationships, once the subject of great paintings, frescos and tapestries, continue to inspire architectural strategy and detail. Large-scale topographical plans, sections and models of landscapes are among the drawings that situate building proposals.

Representation of landscape is itself a major area of study for contemporary visual artists, and architects can certainly draw on the wealth of these studies. At the same time, the focus of the architect's drawing is often not the landscape in itself, but rather a reiteration of an age-old dialogue between nature and artifice; constructed landscapes, rather the landscape as a whole.

It is interesting to note how layering of colours is part of the techniques traditionally used to depict landscapes. As far back as the tempera paintings of the Middle Ages and early Renaissance, for instance, we find woods (nature) were first painted with several layers of black, and then over-painted with thin glazes of the transparent pigment Terre Verde (Earth Green). Perhaps this gaze into deep shadows, representing the mystery of the natural landscape, is echoed in the later great paintings in the layered glazed surfaces of oil and watercolour.

Below
A carefully judged 'digital render' by A E Lee. The ambiguity of this constructed landscape is achieved by balancing render settings of the original model and layering other effects in Photoshop.

Right
This carefully rendered digital drawing by Morphosis represents the structure in its landscape setting, creating the impression of real engagement with the terrain that is important to many architectural proposals.

Case Studies: Landscapes

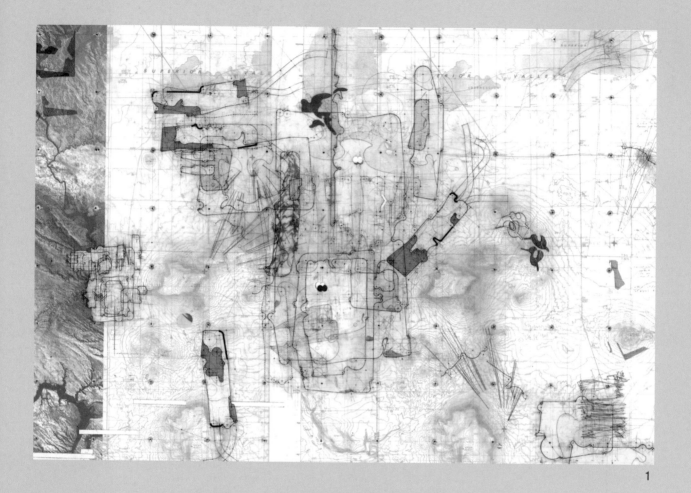

1

1. Perry Kulper's drawing is a particularly rich exploration of landscape at a strategic level. The contoured topography gives rise to a poetic interpretation that appears to be part landscape and part emerging construct. The drawing overlays the plan of the real landscape with a delicate, layered set of lines and tones. This image operates at two levels: first there is an overall arrangement of boundaries and surfaces, drawn in dark pencil on mylar (film) and potential enclosures are emphasized in Naples Yellow, pinks and Sienna Earth; secondly there is an intricate world of detail that negotiates boundaries between the constructed and the explicit landscape, visible through the translucent surface of the mylar. The layering of the drawing in this way expresses a depth of landscape and at the same time presents the image as a vehicle for imaginative interpretation. This kind of drawing is only in part a conveyor of information; it is also a drawing that provokes reflection about landscape and architecture.

2

2. Perry Kulper's exploration of the boundaries between architecture and natural context in plan is illustrated three-dimensionally in the 'digital paintings' of A E Lee. The important shift in Lee's work is that the impact of the landscape is caught as a space; a habitat that is part internal and part external. His approach to drawing is pictorial, carefully adjusting colour balance, brightness and contrast to maintain detail but also to attune shadows in order to maintain depth. The drawing starts with a three-dimensional digital model that is carefully assigned lighting and material. Several renders are done using different settings. After importing these layers into Photoshop, they are overlaid and tested using different effects, blending modes and transparencies.

3

3. Gillian Lambert's drawing of the entrance
elevation for the House at Galleon's Reach is
equally evocative of structure and landscape at a
river's edge. The drawing is an imaginary vision
that is intended to capture the magical qualities
of atmosphere. Studies of wind, captured in
sketches, photographs and line drawings,
were edited and used in the final drawing, both
compositionally and graphically. The entrance
elevation receives southwesterly prevailing winds
and its curtain wall was made of parachute silk.
Layers of pencil drawings and paint applied with
an airbrush were scanned and combined to build
up the blurred and ghostly fabric.

4

4. Stephenson Bell's rendering of their proposal for a visitor and education centre at the Messel Pit, near Frankfurt (a UNESCO World Heritage Site) is equally evocative, and maintains a high level of detail and clarity. As part of the development of the scheme, the architects visited the site and undertook a photographic study with a Canon 350D digital camera. The muted palette of the snow-covered landscape was emphasized by converting the images into black and white, with the contrast and brightness adjusted (so that all the images were consistent). They were then 'stitched' together with Adobe Photoshop to create a continuous background for the computer render of the proposed building. The background images were faded out at the perimeters to blend in with the sky and ground.

The proposal was modelled with 3ds Max and rendered with V-Ray, with the camera angles being set up early to enable the background and foreground to be developed while the model was being created. Textures were taken from existing elements on the site photographs (which were adjusted in Adobe Photoshop to avoid any repetitive tiling). Particular attention was given to the lighting to achieve a soft diffuse light, replicating the winter haze apparent in the site photographs. The ground levels were modelled to form the right shadows when the ground falls away from the projecting walkway, and together with the soft light the appropriate shadows were created.

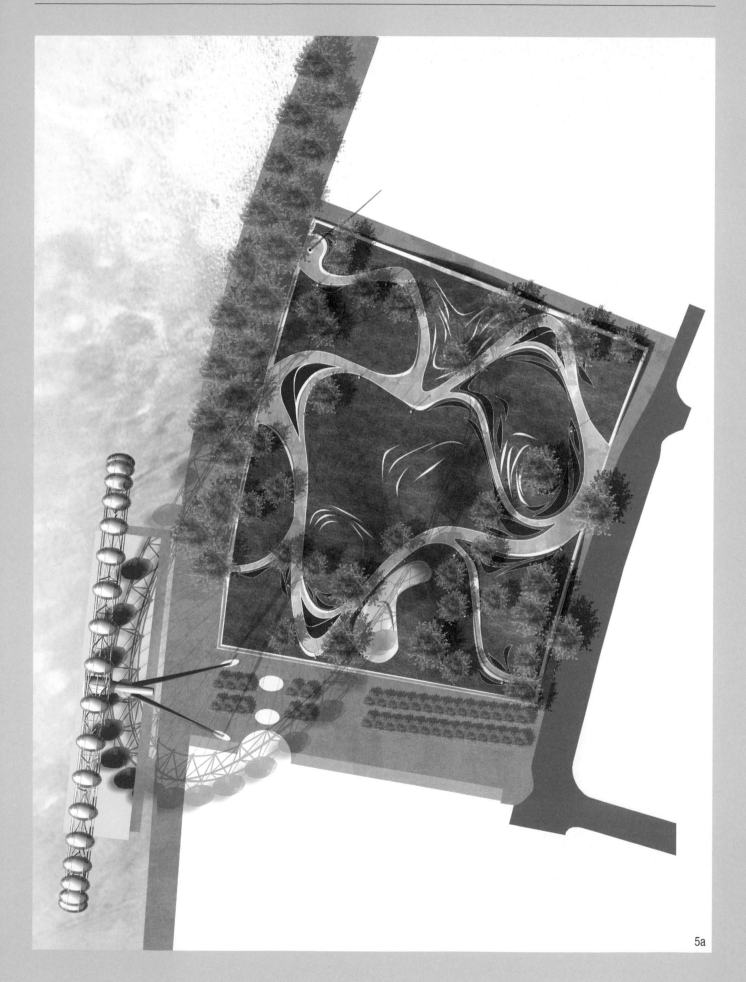

5b

5c

5a, b, c. The plan and view of Queen's Walk (in winter) by West 8 has a similar delicacy in the way that the landscape is rendered. The images combine clarity with a visual impression of the character of the landscape proposal. By contrast there is a more diagrammatic feel to the isometric view (at night) of the overall proposal.

STEP BY STEP DIGITAL PAINTING: LANDSCAPE IN WATERCOLOUR AND PHOTOSHOP

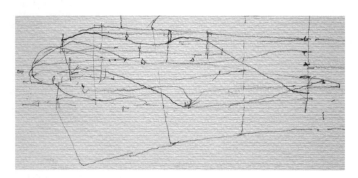

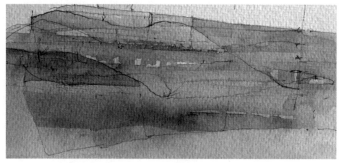

1 This exploratory sketch is based on a landscape study at the early stages of the design process, where the medium of watercolour is a means of thinking through preliminary ideas of location and a response to the flatness and scale of the context. An abstract pencil sketch is not intended to be representative of the specific landscape, but instead fixes certain ideas about measurements and scale.

2 The first washes overlay Burnt Sienna and Raw Umber (natural) onto Payne's Grey to represent a sense of the landscape setting. The washes are loose and are used to find possible ideas for the eventual structure.

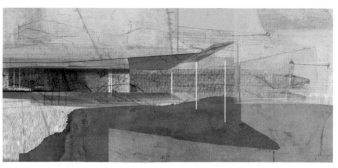

3 Importing the watercolour sketch into Photoshop, a foreground texture is added in a new layer and trimmed using the pen tool. Some of the washes are lightened by wetting the page and then reducing the colour with a sponge.

4 Photoshop is used to complete the sketch image for the new structure in the landscape.

STEP BY STEP CREATING A GARDEN LANDSCAPE USING CAD

A sequence of landscape images created by Sarah Gilby.

1 3D model created in AutoCAD by Sarah Gilby, modelled in AccuRender.

2 New Layer – Sky image overlaid to lighten 3D model sky. Extra layer lightened, opaqued and blended in using the eraser tool (window > tools or short cut E).

New Layers – trees and foliage montaged, edges erased with the blurred eraser tool. Brightness contrast, opacity and colour balance levels adjusted.

3 New Layer. People and objects were added, cutting and pasting from additional images. Garden structure was taken from a photograph of a built model – using the Transform tool to fit it into position. (Edit > Transform > Skew.)

Layers of model were overlaid to add depth.

4 Final image.

TIP SHADOWS

Shadows are particularly effective in three-dimensional drawings of landscapes, particularly of trees and other vegetation.

Urban settings

Urban situations produce a wealth of diverse architectural drawings, varying in scale from strategic studies of a city, its metabolism, topography and culture, to relational studies of more immediate context – the block or the street – and finally through to specific spatial studies that negotiate boundaries, transitions, structure and materiality.

The range and complexity of issues is matched by the diversity of 'urban drawing types' and approaches to design in urban contexts. During the process of urban design, drawings are the vital component in terms of the way they can open up the debate and engage others. 'Urban drawings' are key drivers for the design team, in the eventual resolution of the city's often-conflicting demands. The role of the urban drawing is two-fold: first, to synthesize; second, to communicate.

Urban drawings incorporate wide-ranging specialist inputs and bring together the perspectives of diverse stakeholders. While software can facilitate information management, these complex design decisions still require judgment based on a deep understanding of the city and, like the any other architectural situation, will need to be worked out through drawings, models and artefacts at a range of scales.

One of the dominant kinds of urban drawings today is the perspective, and computer-generated modelling enables building forms to be placed in context with photographic precision. With perspective comes the ability to see the city as a unified scene and also to demonstrate explicit formal relationships. The computer facilitates the 'designing' of the city and allows it to be visualized in three dimensions.

At the same time, however, the expectation of these photorealistic images should not preclude other, perhaps more subtle, studies that foreground less tangible aspects of urban life – drawings that capture, for instance, the performance, metabolism and life of the city.

Left
Anne Desmet's *New Metropolis* incorporates wooden type, wood engraving, lino cut, flexgraph and collage on paper.

TIP CONTEXT

In drawings that show context for a specific building, concentrate on careful observation, measurement and visualization of the immediate surroundings. Detail in further distance matters less.

TIP OBSERVATION

Use drawings and other media over a period of time to develop an understanding of changing conditions and different patterns of inhabitation.

Below

Devanthéry & Lamunière, School of Life Sciences, Lausanne, Switzerland, 2005. This digital render drawing shows the glass entry wall of the building, which opens up the world of research and scientific information to the city. The night scene and transparent foreground figures emphasize the life within the three-storey reception.

Case Studies: Urban settings

1

> **TIP** URBAN SPACE
>
> Use plans, sections and models
> to understand three-dimensional
> patterns of urban space.

1. The first drawing featured here is about simple observation of the city and light by the artist Anne Desmet. The first, *Matera towards Evening*, is a lino-cut printed in blue-black and cream inks on Somerset Satin paper. This lino cut is a direct development of a small, detailed, pencil and grey wash drawing (made on site) of part of the ancient town of Matera, in southern Italy. The sun is just hitting the scene and the drawing emphasizes the way that buildings merge into the shadows; the rocks from which they came. Two printing blocks were used: the key block, printed in blue-black, carried all the structure of the composition; the second block (the same size as the first) was uncut, but inked in a transparent cream tone over the first printing in order to imbue the image with a sense of the yellow warmth of the stone and sunlight in contrast to the bright whiteness of the printing paper.

2. It is interesting how the hand-drawn sketch still remains a key tool with which to synthesize and articulate ideas, even in the complex environment of the contemporary city. This study by Eric Parry shows an urban plan and two elevations for 5 Aldermanbury Square in the heart of London. The pencil sketch establishes, at an early stage, preliminary ideas of strategic importance to the development – in plan, the idea of a generous public space linking urban spaces to either side of the building at ground floor. The scale of this is indicated in the base of the first elevation. On the second elevation another idea, the inward canting of the upper parts of the elevations which is key to the final project, first appears. This device reduces the impression of the scale of the building at street level. In this sense this simple sketch is a synthesis of ideas about the life and experience of the city with strategic and detailed decisions about form, arrangement and structure.

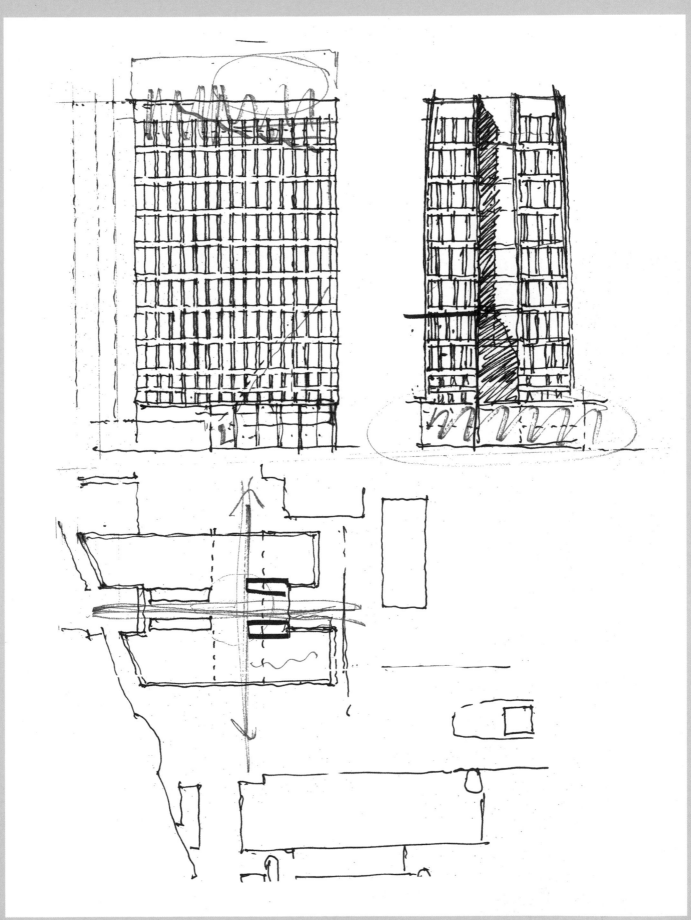

4a

4b

3. Zaha Hadid Architects, CAD image of the Groninger Forum, showing the proposed building in the context of the existing fabric. The effectiveness of the drawing is partly in the way it limits the use of colour to highlight the proposal. The rest of the city is rendered in a blue/grey tone and, while giving the background context, it does not detract from the dramatic building that appears as a light plane cutting diagonally across the image from the bottom right. This composition and carefully-judged play of light and movement are key to the power of the drawing.

4a, b. Zaha Hadid Architects, Kartal-Pendik masterplan, Istanbul (see also pages 102 and 119). This project began by weaving in and out of the existing fabric to create a soft grid that forms the generating plan for a three-dimensional developmental framework (right). Topologies were developed through scripting software that responded to the perceived demands of different districts. The masterplan image (above) was made by taking this adaptable model and rendering it using a range of software including Rhino, Autodesk and Maya.

STEP BY STEP CREATING A PHOTOMONTAGE

In this sequence of images by Ian Henderson a photograph of
an existing urban environment is manipulated to show how a
proposed architectural intervention would look on the site.

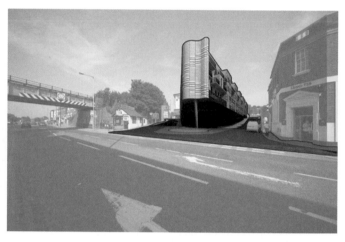

1 Photo. Refine the photograph for best results with colour-correction
techniques using layer or image adjustments such as levels, curves, photo
filter, hue/saturation and so on. Ensure that any unwanted marks and dust
spots are removed using the Clone Stamp and Heal Brush tools. Sharpen if
necessary using either the Unsharp Mask or Smart Sharpen filters.

2 Model. Create a new group and rename. Drag the CG element into the new
group in the working document with the Move tool, holding down shift to
ensure that it lands in the correct location.

Rename the new layer with the CG element. Select the group and apply
a layer mask. Using the Brush tool paint on the layer mask with black and
white to hide and reveal appropriate parts of the building layer so that it sits
well in the photograph.

The CG element was colour-corrected to match the tonal values of the
photograph using layer or image adjustments such as levels, curves,
photo filter and hue/saturation.

3 Background. Work on the background to mask out any unwanted areas
of the photograph. Add trees and bushes to the background using groups,
layers and layer masks. Create a new layer and clone areas of the sky to
cover up the existing building. To blend the new CG road with the existing
road, copy and clone areas of the existing road on to a new layer until the
two elements join correctly.

4 Shadows. To ensure the CG element sits properly in the photograph trace
the shadows cast by the existing buildings using the Quick Mask mode to
paint and create a selection. With the existing shadows traced and selected,
create a new layer and rename. Fill the selected area with black using the
Paint Bucket tool. Set the opacity of the 'shadow' layer to match the value
of the existing shadows.

5 Reflections. Create a new group and rename it. With the Polygon Lasso tool, select the windows. Create a 'layer mask' for the group from the selection. Drag, drop, scale and position images that closely represent what could be reflected in the glass into the new group. Ideally use photographs taken from the other side of the road to provide an accurate reflection; this will require careful planning in advance. Colour-correct the reflection images and use layer blends and opacity values to create convincing reflections.

6 Vegetation. Create a new group and rename it. Drag, drop, scale and position the vegetation into the scene. Colour-correct the vegetation to match the tonal values of the photograph using layer or image adjustments such as levels, curves, photo filter and hue/saturation.

7 People. Create a new group and rename it. Drag, drop, scale and position the people in the scene. Colour-correct the people to match the tonal values of the photograph using layer or image adjustments such as levels, curves, photo filter and hue/saturation.

8 Shadows of people. To create shadows for a person, duplicate the layer. Using the Hue/Saturation Adjustment, create a silhouette from the duplicate layer by setting the 'saturation and lightness' values to -100.

Use the Distort function from the Transform tool to manipulate the silhouette to form a shadow. Take care to ensure that the shadows lie in the same direction as those in the photograph. Move the 'shadow' layer beneath the 'person' layer. Set the opacity of the shadow to match the tonal values of the shadows in the photograph. Use the Gaussian Blur filter to soften the shadow if required.

9 Final image.

STEP BY STEP USING PHOTOMONTAGE AS PART OF THE DESIGN PROCESS 1

These diagrams, drawings and digital models were made as part of the design process for the transformation/renovation of the TSR Tower, Geneva, by Devanthéry & Lamunière.

1 A photograph and two diagrammatic sketches show the existing tower in context and illustrate the three-dimensional intentions as they look to 'open the form' of the existing tower, 'passing from an "I" volume to a "Z" volume, or from a solitary form to a form that merges more interestingly with the environment'.

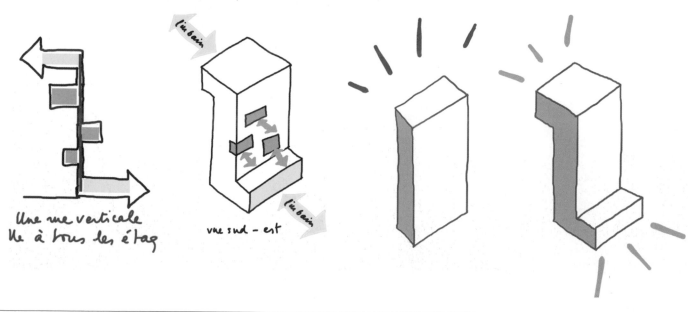

Une vue verticale lie à tous les étag

vue sud – est

2 A computer-generated model.

3 The final contextualized rendered model of the proposal.

STEP BY STEP USING PHOTOMONTAGE AS PART OF THE DESIGN PROCESS 2

These diagrams, drawings and digital models were made as part of the design process for the Lausanne Opera House by Devanthéry & Lamunière. The architects describe this project as entirely oriented on a technique that originates from the magic of the performance. All of the building is clad with a 'skin' of glass and stainless steel that 'symbolizes the brilliance of a nightgown'.

1 Photograph – existing street view.

2 Annotations on urban analysis.

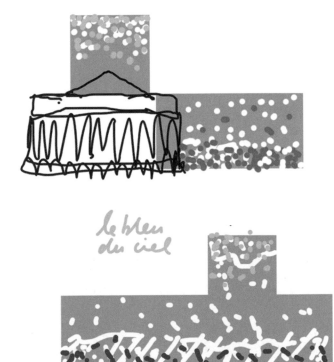

3 Diagrams showing the role of the reflective cladding, accentuating the disappearance of volume at street level and towards the sky.

4 Final rendered CAD model in context, with lighting to accentuate the intentions behind the cladding and form.

Scripted drawings and urban design
Daniel Richards

This section uses algorithmic design processes to explore the organization of urban form. It describes an approach to the design of urban systems and speculates upon ways in which a data-driven design process can be established through the construction of custom scripted design tools. This section is split into three elements which each expand upon the idea of growing urban form from algorithmic processes which respond to various contextual and legislative stimuli.

This case study follows the development of a proposal for Beswick in East Manchester in which emergent architectural and urban traits have evolved in response to existing site conditions and the East Manchester local development plan.

Decentralized networks

The example opposite illustrates how urban infrastructure can be defined using a collection of nodes as design generators and allow local interventions to establish a global network. This network is defined using decentralized control; this means that each node is responsible for creating a piece of surrounding infrastructure based on site-specific (local) relationships

which contribute towards a collective (global) network.

The intention of this case study is not to provide software-specific guidance as to how to write a script, but rather to illustrate the generic steps used to create each script. The 12 diagrams (below right) read horizontally as three iterations of the same process to illustrate how the script can apply differently to various situations.

The script comprises several parts:
1 A population of nodes is created whose positions will define the resulting network.
2 All nodes are collected into an array, allowing them to be accessed via a loop function.
3 A first loop is created which subjects each node to the following operations:
    4 All other nodes within X distance of the selected node are defined as immediate neighbours and are added to a second array.
    5 Vectors are projected between the selected nodes and immediate neighbours and the mid-point coordinates are stored.
    6 The mid-point coordinates are sorted by their angle around the selected node and a spline is drawn connecting the sorted mid-points.

The infrastructure that emerges becomes an interference pattern of the initial node positions, thus creating an infrastructure that has grown out of a simple construction algorithm and the local relationships between the nodes. In addition to this example, voronoi diagrams can also be used to construct optimum networks between a set of points and have been used extensively in urban planning establish zoning and catchment areas.

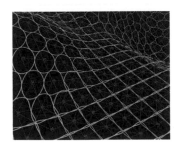

Establish region and find neighbours

Collect mid-points

Generate connections

Emergent infrastructure

Establish region and find neighbours

Collect mid-points

Generate connections

Emergent infrastructure

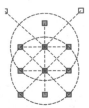

Establish region and find neighbours

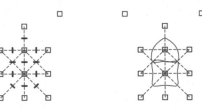

Collect mid-points

Generate connections

Emergent infrastructure

Below
Network patterns and associated
parametric equations.

Opposite right
Network growth through a simulated landscape.

Below are 12 emergent urban networks which are defined by their associated parametric equations that have been, in turn, used to distribute the control nodes. The urban networks all exhibit emergent traits: village greens are created when nodes are arranged in a circular formation [3]; large inner ring roads form in radial networks [4] and suburban areas with decreasing density can been seen on the fringes of most networks. Importantly, complexity has not been given to these networks; their complexity has developed as an emergent property of local interventions.

The script that was used to create the urban networks can also be extended to construct three-dimensional structures. The example below shows a similar process in which a population of nodes create three-dimensional surfaces between neighbouring nodes. In the example a secondary script has then been applied to subdivide the surfaces and construct curved elements.

The process of decentralizing the design processes allows for the creation of multiple scripts that can be run sequentially to produce paradoxically controllable, 'out of control' structures.

Growing networks

The previous explorations illustrated how algorithmic design can benefit from embracing decentralized control.

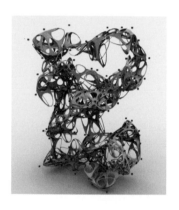

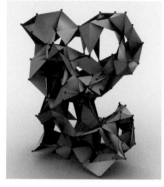

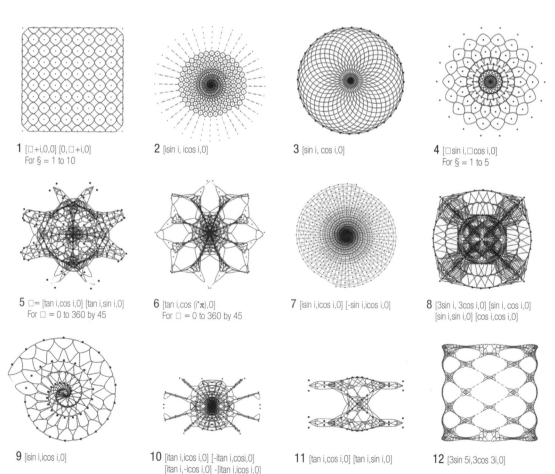

1 [□+i,0,0] [0,□+i,0]
For § = 1 to 10

2 [isin i, icos i,0]

3 [sin i, cos i,0]

4 [□sin i,□cos i,0]
For § = 1 to 5

5 □= [tan i,cos i,0] [tan i,sin i,0]
For □ = 0 to 360 by 45

6 [tan i,cos (i*π),0]
For □ = 0 to 360 by 45

7 [isin i,icos i,0] [-sin i,icos i,0]

8 [3sin i, 3cos i,0] [sin i, cos i,0]
[sin i,sin i,0] [cos i,cos i,0]

9 [isin i,icos i,0]

10 [itan i,icos i,0] [-itan i,cosi,0]
[itan i,-icos i,0] -[itan i,icos i,0]

11 [tan i,cos i,0] [tan i,sin i,0]

12 [3sin 5i,3cos 3i,0]

This example will follow the creation of a script which can grow road networks through a simulated urban landscape.

In 1970, John Conway devised a zero-player game called the 'Game of Life', is a mathematical 'game' that simulates colonies that grow and perish over time. The game consists of a two-dimensional grid that defines a series of cells. The cells can exist in one of two states: alive or dead and each cells state is defined by its relationship to its neighbouring cells. The game then works by iterating very simple rules in which each cell must evaluate its neighbouring cells and define the current state. Over generations, the game is able to simulate organic behaviour. Conway's 'Game of Life' is the most well known example of a computational technique called cellular automata.

The script used to grow networks requires two elements. First, a landscape must be established and second, a generator for growth must be designed. The landscape becomes defined as a grid of cells (as Conway's model), providing a space in which networks will be grown and also a mechanism by which cells can communicate using decentralized controls. The generator for growth will be based on the biological phenomenon 'Phototropism' which is the tendency for plants to grow towards sources of light. The aim of this script is to establish a cell as the goal point (light source) and a process by which suitable

transport routes can be grown through the pixelated landscape to provide a network link.

The script comprises several parts (as shown in the diagram below). Each cell can exist in one of two states: open space or existing building.

1 Each cell has the capacity to communicate with its immediate neighbours only.

2 A target cell is identified.

3 Using an iterative function each cell connects to the neighbouring cell closest to the target cell until the target is reached.

4 Cells which are defined as existing buildings form obstacles. Subsequently the next shortest route may need to be taken causing a detour.

5 Cells connections to neighbouring cells may not always form a connecting link on first trial. On such an occurrence the network reaches a dead end surrounded by existing buildings.

6 As cells connect to neighbouring cells they drop 'digital breadcrumbs' that can be retraced upon meeting a dead end. These 'breadcrumbs' are stored within a small memory bank within each cell that has been created by assigning each cell a custom attribute.

7 Agents are now able to find their way out of dead ends and dead ends are then filled in to prevent future networks making the same mistake.

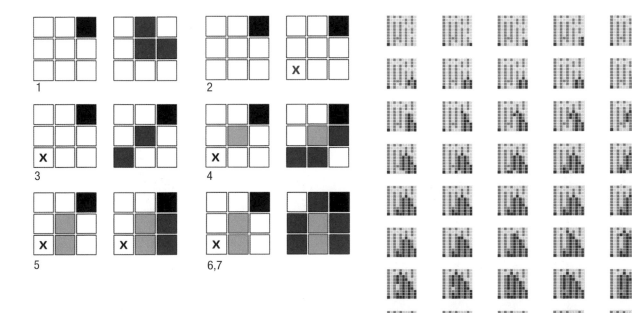

The network growth script works by growing connecting networks between assigned target cells and starting cells. The script provides connectivity through any landscape where connection is possible.

The example, below left, illustrates a network being grown through a randomly generated landscape whereby each cell on the top and left sides of the grid are connected to a target cell in the bottom left corner.

The network script can also be formatted to operate within a three-dimensional landscape. The example, below right, shows the emergent network of routes through a three-dimensional maze. The physical model has been printed in gypsum powder using a Zcorp 3D printer.

The resultant networks are highly rational and theoretically predictable. However, the sheer magnitude of calculations required to manually compute these networks make it implausible to complete without the use of computation.

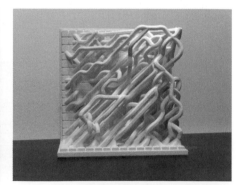

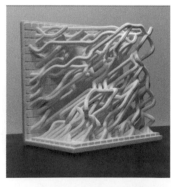

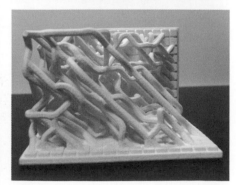

Evolving networks

The process of writing scripts to generate form is not difficult. However, for scripted methodologies to become 'useful' it is important to develop methods by which resultant forms can be processed and evaluated.

The example, below, shows how the networks grown using the previous script can be subjected to evolutionary pressures using a genetic algorithm to simulate Darwinian natural selection and thus breed fitter urban networks.

The network landscape has been created as a grid where each cell is either open space or existing building, therefore the landscape can be reformatted and stored as binary code with 1 referring to existing building and 0 representing open space. This process requires a function which can examine any landscape and print the landscapes' genetic code to a .txt file which can be used later to upload landscapes and thus associated network phenotypes from binary.

Once the landscapes and thus resulting networks exist as pure binary data they can be easily be subjected to

'genetic crossover' to create hybrid urban models which exhibit traits from two or more parent landscapes/networks.

In order to provide points at which beneficial mutations may occur, the potential for copying errors must be introduced to the genetic crossover.

Once the mechanisms for downloading, uploading and crossover with potential mutations have been established we can develop an organic design methodology using a genetic algorithm to evaluate and evolve the network models.

1 Download landscape

2 Upload landscape

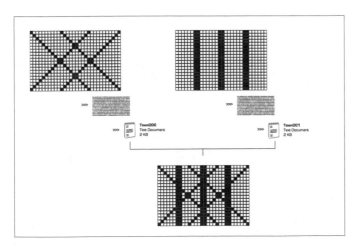

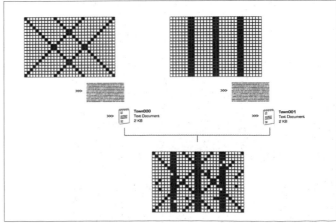

3 Genetic crossover

4 Potential mutations

The diagram, below, illustrates the algorithmic process of evolution. Beginning with an existing urban landscape (Beswick, East Manchester) we download the landscape to binary and upload three offspring landscapes which have been subjected to possible mutation. The network script is then run on each landscape producing a fully-connected infrastructure and defining certain cells-specific programmes based upon their local relationships. Each networked landscape can then be evaluated for variable features such as: transport route lengths or the amount of new residential space (red cells) allowing each offspring landscape to be assigned a fitness level.

Over a series of iterations, the process of using the 'best' landscapes/networks to breed potentially improved versions of the original urban space can lead to optimized urban form with complex emergent characteristics.

The design of urban space using algorithmic tools enables the designer to establish soft topological relationships between spaces, rather than definite physical connections. The act of drawing becomes augmented. Sketching becomes the act of strategizing relationships in code and technical drawing transforms into the writing of architectural scripts used to develop bottom up generative systems.

Below
Phylogenetic tree of successful urban networks.

Opposite
Emergent architectural traits and distribution of programmatic elements.

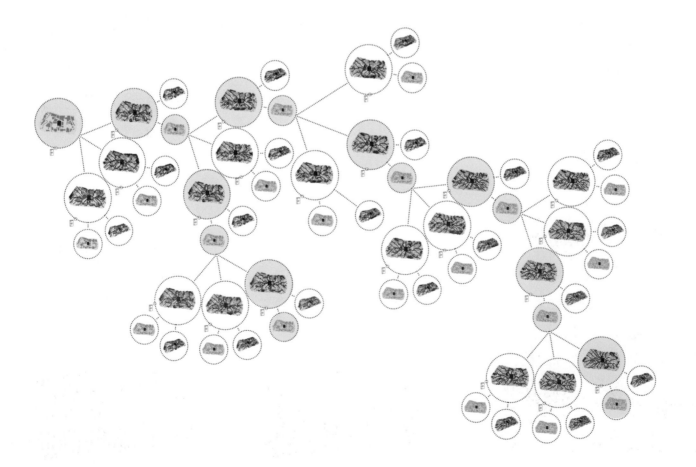

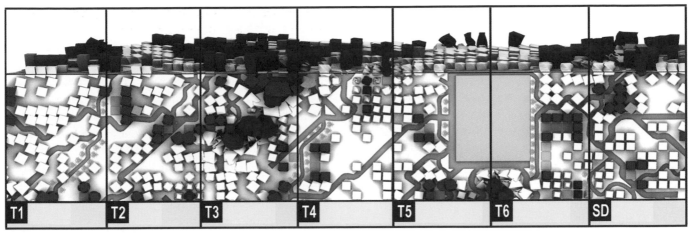

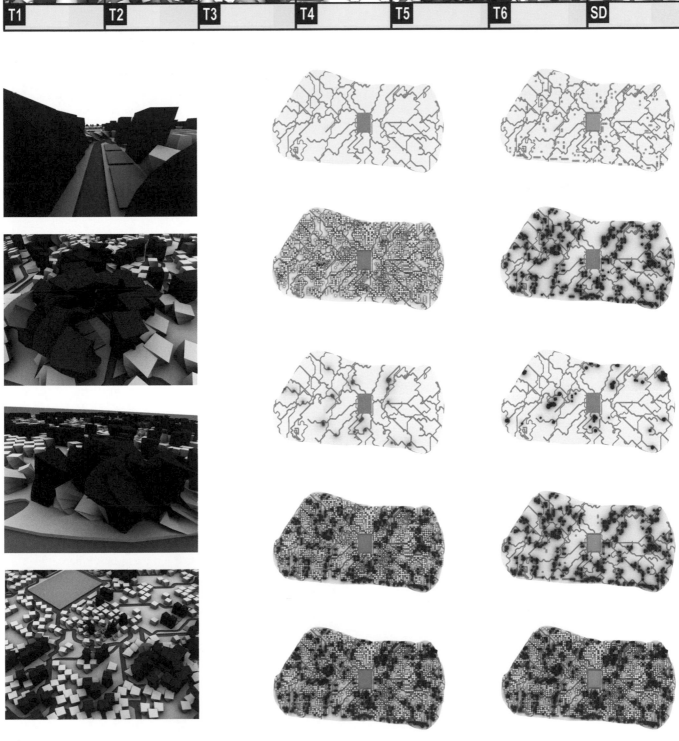

Glossary

There are a number of key words relating to drawing that can be interpreted in different ways. It may be useful here to provide a broad definition of the more important terms used in this book.

Axonometric

Axonometric projection is one of the most frequently used drawing types to create three-dimensional images in architecture. Isometric, dimetric and trimetric projections are types of axonometric drawings and together they form a group of drawing types called parallel projections or 'paraline drawings'. Axonometric is the term used in architecture to describe a projection from a plan drawn to scale: plan elements are projected vertically and to the same scale as the plan, which is first rotated to provide the intended view. Walls that appear at the front of the plan can be omitted or only partially projected to create a 'cutaway' axonometric, that reveals the interior that would otherwise be obscured with uniform projection.

Collage

Collage techniques can be used in architecture to visualize interiors, exteriors or, most often, to represent a building proposal in a new context. In their simplest form they are cut or torn images or drawings glued onto a surface. More often they are done in Photoshop, using layers to visualize different images in the same plane. Collages can also be used to illustrate a design proposal, perhaps exploring context, light and materiality but also as a key device in the development of a design.

Elevation

Elevations are orthogonal drawings, usually parallel to the surface in question (which may be internal or external). Elevations are like vertical sections of a building that illustrate wall articulation frontally. Architects use the term as meaning the same as facade. An elevational drawing may be effectively used to define relief using shadows, colour and textures.

Isometric

Isometric projections are types of axonometric drawings and together with isometric and dimetric drawings form a group of drawing types called parallel projections or 'paraline drawings'. Isometric drawings are formed when the object is turned so that all three axes make the same angle with the picture plane, making the angles between the edges of the building or space 120 degrees. In architectural drawings one axis is usually vertical and the other two therefore at 30 degrees to

the horizontal. Most architectural drawings are shown looking down, (although details and ceilings may be best shown looking up) and view a plan that appears to be 'opened up' in the direction of viewing from 90 to 120 degrees. This has the advantage if the drawing reveals an interior, but otherwise is a relatively rigid projection that requires all three visible planes to be emphasized equally.

Linocut

A linocut is a simple but effective form of relief printing. First the drawing is transferred onto a piece of linoleum (lino). The lino is then carved to take away those areas that are not to appear in the final print. It is then inked with a roller or brush. Even pressure is applied to a piece of paper placed on top of the carved lino (by using a printing press or burnisher) thus transferring the ink onto the paper and creating the print.

Monoprints

The simplest form of monoprinting, called trace printing, is a useful technique for architects: ink is first spread evenly onto a surface and then a piece of paper with the drawing printed in reverse is then laid (face up) onto the ink. By re-applying pressure over the lines of the reversed drawing, that same drawing will be transferred onto the other face of the paper, but the correct way around. This simple process gives a more individual quality to a line drawing.

Orthogonal

Term used to describe a drawing or projection that is made up of or involves right angles.

Perspective

The technique of perspective drawing is a representational tool to depict distance, or spatial depth on a two-dimensional plane. Today the persuasive power of mathematical perspective persists: the computer-generated perspective has become the singular most powerful tool to communicate a project. There are two fundamental observations embodied in perspective. Firstly, that objects appear smaller in the distance than they do close up. Secondly, that objects appear to become foreshortened along the line of sight. Only two kinds of perspective – single-point and two-point perspective – are described in this book, but the number of vanishing points in any perspective will depend on the different orientations of the configuration.

Photomontage

This a collage constructed from photographic images.

Pigment

Pigment is the substance or powder that, once dissolved in a medium, makes up the colour of paint. The magical character of pigments is their transparency: each pigment varies in the degree to which it is transparent. The basis of traditional rendering techniques is an understanding that depth of colour is created from layers of transparent glazes, laid one over the other. In architecture we see this most often in watercolours, but raw pigments are flexible and it can be interesting to work with them directly, using other mediums.

Plan

A plan is a fundamental architectural drawing type. It is a primary organizing device and as a consequence it is the central drawing to a great many projects, and is the drawing through which buildings can be most easily read. Technically a plan may be described as an orthogonal projection from the position of a horizontal plane. The position of this plane and the scale of the drawing can produce a wide variety of plan types, ranging from landscapes to details.

Scale

The standard metric units used on most architectural drawings are the (SI) units millimetres (mm) and metres (m). In France centimetres (cm) and metres are most often used and in the United States drawings are also done to non-metric scales.

Common metric drawing scales for are 1:1 (e.g. full size fabrication drawings); 1:5 (e.g. construction drawings); 1:20/1:10 (e.g. technical sections); 1:50 (e.g. detail plans or sections); 1:100 (e.g. arrangement plans, sections); 1:200/1:500 (e.g. overall arrangement drawings) and 1:1250/1:2500 (e.g. context drawings).

Sciagraphy

This is a technique of projecting shadows in drawings and is useful to illustrate surface relief and spatial depth.

Scripted drawing

Scripted drawings are images that are produced by so-called 'generative software'. These drawings represent formal solutions to complex design parameters, understood mathematically, where the design process can be broken down into an algorithm, or set of instructions.

Section

Like a plan, a section is an orthogonal projection, but from the position of a vertical plane. Sections can be cut anywhere through a building, but tend to be taken where there is a significant spatial condition to describe. Internal elevations will appear between floor plates and an external elevation appears in a section where the vertical plane is taken from outside, or partially outside, of the building.

Further reading

Berger, J., *Berger on Drawing*, Occasional Press,
Aghabullogue, 2005

Ching, F., *Architectural Graphics*, John Wiley & Sons,
New York, 2003

Cook, P., *Drawing: The Motive Force of Architecture
(Architectural Design Primer)*, John Wiley & Sons,
New York, 2008

Cooper, D., *Drawing and Perceiving: Real-world Drawing
for Students of Architecture and Design*, John Wiley & Sons,
New York, 2007

Edwards, B., *Understanding Architecture Through Drawing*,
Taylor & Francis, London, 2008

Giddings, B., and Horne, M., *Artists' Impressions in
Architectural Design*, Spon Press, London, 2002

Holmes, J. M., *Sciagraphy*, Pitman, London, 1952

Leach, S. D., *Photographic Perspective Drawing Techniques*,
McGraw-Hill, London, 1990

Martin, L., *Architectural Graphics*, Macmillan, London, 1970

Porter, T., and Goodman, S., *Design Drawing Techniques:
For Architects, Graphic Designers and Artists*, Butterworth
Architecture, Oxford, 1991

Porter, T., *The Architect's Eye: Visualization and Depiction of
Space in Architecture*, Spon Press, London, 1997

Reekie, F., and McCarthy, T., *Architectural Drawing*,
Edward Arnold, London, 1995

Uddin, M. Saleh, *Hybrid Drawing Techniques for Architects
and Designers*, John Wiley & Sons, New York, 1999

Vesely, D., 'Architecture and the Poetics of Representation' in
Daidalos, No. 25, 15 September 1987, Seductive Drawings,
pp. 24–36

Yee, R., *Architectural Drawing: A Visual Compendium of Types
and Methods*, John Wiley & Sons, New York, 2007

Index

Page numbers in *italics* refer to
picture captions

Picture credits

Author's acknowledgements

For Francesco

I am grateful to all those who have contributed to this book. Particular gratitude is due to Ian Henderson, Helen Murgatroyd, Perry Kulper, Chris Staniowski, Sara Shafiei, Ben Cowd, Matthew David Ault, Daniel Richards and Anne Desmet for their expertise, inspiration and significant contribution; to Gary Butler, Sarah Gilby, Kyle Henderson, Sophie Mitchell, Phil Meadowcroft, Eric Parry, Mario Ricci, Peter Sparks, and Janek Ozmin, Gillian Lambert and Kathryn Timms whose work forms a large section of the book. Thanks also to Autodesk for their collaboration, to Yvonne Baum and Gemma Barton for their work at the early stages and to Nick Dunn for his collaboration and support. Finally thanks to Philip Cooper and Liz Faber and the team at Laurence King Publishing, and to editor Mark Fletcher and designer Vanessa Green for their enormous support, patience and hard work that made this publication possible.